π
BASIC MANUAL OF PYRAMIDOLOGY

Gabriel Silva

BASIC MANUAL OF PYRAMIDOLOGY
Gabriel Silva –
Second Edition - Edit. March 2016
ISBN 9781326593223
All Copyright reserved.
More info: www.vitalpyramid.com
www.piramicasa.es (Spanish, more complete about pyramids)

BASIC MANUAL OF PYRAMIDOLOGY

Index of Basic Manual of Pyramidology

Chapter I - Declaration of Intentions
Chapter II - For What Pyramids Are Used and How Is Used
Chapter III - Brief History of The Pyramidology
Chapter IV - Pyramids and Antipyrámids
Chapter V - General Guidelines of Pyramidal Therapy
Chapter VI - How to Make Pyramids
Chapter VII - The FAQs
Cap VIII - Fundamentals of Pyramid Effect
Chapter IX - Experiments with Pyramid
Chapter X – Therapy with Anti-Pyramid
Chapter XI - Friends And Enemies of Pyramidology
Chapter XII – The Maximum Symbol of Health.
Chapter XIII – Pyramids of the World .

NOTE OF AUTHOR: We refer often to the "perfect pyramid". This is the one that is built with the appropriate proportions and materials, implementation in correct orientation to produce the pyramid effect. But all of our products have the same anti-lithiasic, anti-aging, anti-oxidants, bacteriostaric, anti-rheumatic and other properties.

We write these things and devote our lives to researching the power of the pyramids, but now the challenge is to deliver this scientific advance to Humanity. If you want to make your own experiments, you will find the correct and updated information. I led a team of quantum physicists (1984-1991), using a mathematical elite. We did over a thousand physical, chemical and biological analyzes ... We obtained wonderful results (as cure my severe rheumatism), made discoveries of historical importance. We know that the Nobel Prizes are not reliable now. (There are winners who are engaged in the war).

Henry Ford said about usurers who did not trust their cars:
"You, maybe not walk in my car, but your children, who will use insurance. And your grandchildren do not conceive the life without my cars".

I say the same about the pyramids dedicate myself to make.

Gabriel Silva - March 2016

Chapter I
DECLARATION OF INTENTIONS

The pyramids never ared tombs. Actually they are the best medicine for rheumatism, and other diseases. Our special bed anti rheumatic is one of many pyramidal products.

Great part of the exciting mystery of the pyramids is resolved long time ago. I am convinced that as well as the wheel represents a major milestone in the technological development of a civilization, then through the domain of the electricity, the ability to fly, or travelling the seafloor and space, the pyramid represents a fundamental tool, but not only in the technological evolution, but a step higher in relation to the understanding of the universe and the development of consciousness. Therefore, for those of us who have for decades of work in them, it seems incredible that such absurd theories persist and a huge disinformation on the scientific results obtained by teams from around the world. However, this is perfectly understandable and I will try to clarify it.

Since a couple of centuries ago the Egyptian and Mayan pyramids have been capturing the attention of people around the world, being for the majority a simple tourist attraction, but also a battlefield between alleged "sceptical" and various investigators, which generates a confusion increased with matters Para psychological and mystics misinterpreted by the misinformation also prevailing in these fields.

This document is to open a door to the knowledge they should modify the criteria of all humanity with respect to the pyramids, its history and its applications, tease out once and for all the public reader, the key points, so that you can satisfy your intellectual curiosity, at the same time to learn how to take advantage of the outcome of the work of the true skeptics, that - far from discuss to maintain positions- we have investigated interdisciplinary and practically, from both the historical and archaeological review, in the labs. I am aware of the difficulties that according to their beliefs and education can take the reader into accepting certain assertions - mandatory in these explanations-, but appealed to their analytical ability, common sense and above all, their curiosity and sense of practicality, since entering the wonderful world of the pyramid experience has cost close to "zero". Just manufacture some cardboard pyramids with the elements detailed below, to make their own experiments, to prove to himself and to anyone who wants to see it, that there is a pyramidal effect undeniable, material, nothing imaginary and not at all from the faith or the suggestion. Will accept or not the explanation of the physical - and there are some thanks to deep research practices, rather than theoretical - the pyramid effect is already certified by medical institutions in Cuba, where these effects are applied - as in Spain and several more countries - in therapeutic utilities from a few years ago, with more than thirty thousand people treated successfully in the treatment of different diseases, as well as very significant uses in farming, beekeeping, animal husbandry and veterinary.

I hope with all my heart that the readers enjoy the pyramids as i do since 1965, when I started - still a child - to become obsessed by discovering the truth about these great works that left us the ancient scholars. In 1973 the pleasure of making my first pyramid was increased with the experimental successes, not as clear as you will have the reader, because he will have with knowledge that I was not at that time. In 1984 the pyramid allowed me to continue living thanks to its healing power, for which he also defined my responsibility for life. In 2000, after trying in vain, for years to find someone able to mount a good business that will allow people to enjoy the scientific advances achieved, I found myself in the dilemma of forget about the matter and take advantage of the knowledge for my exclusive benefit or leaving my comfortable professional status to accept the challenge imposed by the conscience.

In ten years I didn't find any builders, carpenters or entrepreneurs willing to gamble by a product as "controversial" as a house, a bed or any product pyramid, although they offer dozens of people all over the scientific material, drawings, designs and advice for their commercial exploitation. I could write long anecdotes on the matter, with incredible nuances, dramatic and even comedians of the responses received. But when I was about to leave the matter to dedicate myself to take advantage of other talents and of the pyramids in a strictly personal use and family, serving socially as I could in other fields, I had the satisfaction of indescribable learn that Dr. Ulises Sosa Salinas was using pyramids to heal people in Cuba. From the first contact sprang a friendship that soon became true brotherhood, because we share the same ideal, the same love for humanity, the same feeling of universal respect for, the pyramid being the most important factor in common scientific interest. After a few days of deep meditation take the decision to face my own fears and limitations, against social prejudices, against the flouting of the false sceptical and I began to review the experience and all the material accumulated in three decades. If there was a doctor capable of tossing his prestige, his qualifications and deal with the entire world to contribute something so important to the society, I could not live in the shame of having spent all my goods purchased and inheritance in the physical and causal investigation of the pyramid effect, to finally stay with an exclusive use.

Therefore I declare that my intention, shared by many people already brave and wise, as the Dr. Ulysses Sosa Salinas, a pioneer of the pyramid therapy in Cuba, is to provide to humanity the knowledge collected so far. The Homo sapiens sapiens has not yet been won that classification so proud of doubly sapiens, because you can recognize toward Transcendence, is causing their overall self-destruct. A true civilization is one that has undergone the concern by the subsistence, to take care of its significance. The ancient builders of pyramids were not exterminated, massacred or became extinct. We are one step away from unlocking the mysteries of his disappearance and everything indicates that the pyramids who left have to do with a very different destination to which assumes officially. This book aims to bring the reader to these matters without falling into pseudo-scientism or extravagant speculation, but based on the knowledge currently deducted and applied that anyone can check.

Chapter II
FOR WHAT PYRAMIDS ARE USED AND HOW IS USED
Rheumatism is not hopeless

In China, in the region of Xi'an, there are dozens of pyramids. One of them is 365 meters on each side of base and can be viewed using Google Earth, in 34° 23' 10" N and 109o 15' 13" E. throughout China there are about eight hundred pyramids, many of which exceed in size to the Great Pyramid of Giza. In Central America continue to be discovered pyramids, totalling close to seventy and has barely been the era of discoveries in South America, where there are also some dozens. In Bosnia has been found which may be the world's largest pyramid, rivalling with the Large White Pyramid of China, which would be about 500 meters on each side, but is in a site where the satellite public photos do not have sharpness.

All of them were awarded quality of tombs, although only an Egyptian, the small pyramid of Titi, could have been used as a tomb but not by their builders, but millennia later. But we have been discovering the real uses of the pyramids, thanks to the pioneers who mentioned in the next chapter. The utilities that the builders were given to the pyramids were in line with the profound scientific and technological knowledge that allowed them to carry out similar works that even today are no longer shocked to engineers, architects, mathematicians, astronomers, petro logos, masons, builders, and technologists in various specialities. In the book "Sacred Technology of the Pyramids", edited in http://www.lulu.com/spotlight/piramicasa, addressed the topic very widely and the abundant archaeological arguments against the idea of aberrant the pyramids as tombs of kings and pharaohs allegedly megalomaniac.

These work was made by people in previous eras to the predynastic, they are the largest and most perfect made so far, by its various utilities and considering in some cases, such as those at Giza, the greater part of utilities simultaneously.

A synthetic enumeration in increasing order of importance is the following.

1.- Retrieve the edge of the razor blades. There is no abrasive effect, what happens is the crystalline restructuring and elimination of the pore water. Finally the razor blades lose their utility by wear, but having served between 20 and 30 times more. A good sheet, which will allow shave five times, lasting at least more than one hundred shaved. Virtually nothing is oxidized by effect of the air inside the pyramids, except to make quick reactions and specific. OR is that the spontaneous natural oxidation is almost zero in most of the volume occupied by the pyramid.

2.- Conservation and revitalization of the water, which is not bactericide, becomes more wetting and has therapeutic applications. When it becomes more known the enormous amount of curious characteristics of the water, this point will become much more important in both the scientific as well as their applications.

3.- Indefinite retention and revitalization of food. For example, a goat yogurt without any class of preservatives, is left in the pyramid on 10 August 2005, is the day February 03 2007 and is in such conditions that ... I can't resist and it took me half of the bottle with all the pleasure that implies excellent yogurt of the Catalan people de Segur (one of the best in the world). The temperatures in that room all summer it's like the outside environment: Up to +38 °C. At the winter, the heating of the house doesn't get lower than +15 °C. Twenty-four hours later I am still alive, writing this, and thinking in dispatch me the rest, as if it takes a year and a half and seems to be newly developed, I do not think it can decompose while follow under the pyramid effect. It is sometimes done as curd, separating the colloids of the plasma. Six months ago I ate the contents of one of the bottles and was better than good, so I also take the plasma. Yesterday - because only remains for me bottle and a half - were homogeneous. Years ago, there was experienced by up to three months, but this time - all in the science, I have refrained gluttony a long time.

[Updated this information on 12 January 2011 because today I drank half the delicious bottle that I was.] The vegetables and fruits become dehydrated without starting the process of putrefaction, so once treated and packaged with appropriate systems, can be kept indefinitely.

The acidic fruits tend to lose acidity and can be checked that are sweeter. I haven't eaten beef left about 32 ºC in a pyramid during 42 days, without any type of additive, not even mesh network to prevent the action of the flies. These do not make their stools within the pyramid when it has good density. Their instinct tells them that their offspring will have no food, although the meat this tantalizingly fresh because there is no spoilage and the larvae would die of starvation. Even in samples with larvae of flies left in the pyramid, you can check that these migrating hastily, looking for exit of the pyramid.

The dried meat, in that case, she resumed her status as just been cut from the cow a while after, to leave it on the natural stone countertop. Other times has not been reconstituted its hydration, having completely mummified, but has always maintained its flavour and aroma of fresh meat.

USES OF PYRAMIDS

4.- Treatment of fluids in general. In spirits, wines, oils and other products, supports the use of ferromagnetic, especially when combined in the devices the use of pyramid, ant pyramid and the external field of both. The cellar of Montegaredo pyramidal, Boada of Roa, near Ribera de Duero, is a sample of the wisdom pyramid applied by its owner. Its wines have not only the quality of its own in the region and the crafts of the development, but that its fermentation in a pyramid where the pots vineries are in the middle of the plane of the base, occupying space of the pyramid and the ant pyramid alike, prevents the action of any putrefactions, allowing the symbiotic bacterias of the wine to do its job. It then makes the upbringing in the underground reservoir adjacent, taking advantage of the external pyramid field. All of this leads to an unsurpassed quality, even for the most demanding tasters.

While any good wine improves in the pyramid, the artificial wines - the majority of the wines very cheap- break down up to not being able to drink. In fact regain their true state, distinct its chemical components. Some artificial liquors manufactured with good products, improve to resemble the best artisanal liqueurs. The oils deserve a separate chapter, as its constitution requires molecular paramagnetic materials to achieve better grades, in both conservation and production. However, some oils can be treated with pyramids of any material.

5.- Indefinite retention of germination. Any seed that will leave them in a perfect pyramid may be kept valid indefinitely. The maximum time that I have experienced has been five years for a sample of seeds of carrots, which normally expire in a couple of years or at most three. More than three hundred seeds are sprouted up all, except three in the account, possibly by planting them in the same pot.

Carrots have exceeded the normal average size for the variety. In Cuba has been made more methodical statistical work in this application, with a substantial increase in grain production. All seeds can preserve their germinative power and all qualities for an indefinite period.

6.- In beekeeping manages to improve performance, while avoiding the bees many of the diseases, including the fearsome fungus and disease Nosema disease). The larva plastered. Treatment with ant pyramid lets you delete it in session of half an hour, with two hours of interval and a total of four applications; i.e. in a day of application. This is veterinary in all kinds of animals with the same results as in human beings, so that the placebo effect that some assume, has nothing to do with the pyramid effect.

7.- Correction or cancellation of geopathogenic points. The geobiologist still have many reservations about the pyramids, only by its misinformation regarding the same and the results achieved so far in the research and application. However is achieved with pyramids of a certain size -as the Pyramid-house- undo the effect of the point's geopathogenic of small and medium power, since the pyramid field moves the lines of Hartmann and Curry interfere without modifying them. In Ukraine has been achieved override geopathies very large and it has been decontaminated water tables, using pyramids of large copper (dangerous material to remain inside, but useful in these applications),

8.- The most important current applications are the therapeutic dose. The pyramids are used to cure some diseases and/or prevent its onset; especially those resulting from molecular malformation by any cause, such as the rheumatic, mielitic diseases and other of unknown aetiology and considered incurable by allopathic medicine (has been always tried with total success multiple sclerosis, amyotrophic lateral sclerosis (ALS) and ailments clinging as enclosing, bursitis and other areas where the bones are welded. The process takes months or years, with gradual improvement but constant, using pyramids powerful. It has been successfully combated various diseases with bacterial aetiology, since nothing was rotting in the pyramids and this prevents the formation of colonies of saprophytes, while not interfering in any way with the symbiotic bacteria responsible for the digestion, or prevents the function of the phage responsible for cleaning the body of foreign organisms and parts of dead cells.

Sleep inside a pyramid oriented properly and built with the appropriate materials (always paramagnetic as the high-purity aluminium, wood or glass), represents the near-impossibility of rheumatic diseases, bacterial infections or degenerative. Do not yet know to what extent prevents viral diseases, but we have had evidence that these very difficult to attack an organism whose water molecules are well structured, the immune system in perfect condition and marked deficiency of bacteria lysed. Viruses need bacteria and dead cells to draw on the Bahai gene material. Improvements in the immune system have been apparent in many patients treated with pyramids. Interestingly enough, there is a case of kidney transplant, in which the improvement of the immune system has not provoked the rejection, but if you have activated your own kidneys that didn't work. Or is that the patient now has three kidneys working.

The Pyramid is perfect medicine for rehuma, the special bed anti rheumatism is desingned for complet piramidal effect. The relaxation is only one more.

The effect of relaxation achieved has finished in all the cases reported or that have been followed, with sleep disorders, providing some a great clarity in thedreamlike memory (remember perfectly the dreams), without preventing achieve a deep rest. In others, who were taking medication in order to be able to sleep a little, have gradually left the pads, because their sleep is not very deep but remember nothing of their dreams. There are some myths in relation to the paranormal developments that can be achieved by sleeping in a pyramid, but this does not generally happen. What happens is that the people that have a potential paranormal close to its development, reach easily if they wish, thanks to a better overall health, optimum organic functionality and stability of their brain rhythms. The pyramid does not stimulate the paranormal developments so spontaneous, so that requires the attitude of the volitional subject. There is no implicit psychological action (except in dreams) so it is not to expect more than a trend of "catharsis" purification or psychological, to a large extent due to the remedial action.

Hercules Model: fot tratment of rheumatism, is a quantic medicine. You can be can be treated multiple sclerosis or infections tank to anti bacteria power. People hopeless for diverse diseasses find solution.

Cluster headaches, migraines, headaches by electromagnetic saturation and overall functional disorders of unknown aetiology, variable or questionable have been treated in the pyramid, they have submitted fully in all the cases treated up to now.

Some have required prolonged stay or combination of therapy and pyramidal ant pyramidal (as in some rheumatic problems). The osteoarticular problems in general have been the most discussed until now, and in the book "Therapeutic Revolution of the Pyramids", which I wrote in co authorship with Dr. Ulysses Sosa Salinas, plasma is a summary of research protocols and statistical application in Cuba.

A clear example and quick action of the pyramid is the ankle sprain, because the proper pyramid therapy allows the patient fully healed in two days, rather than remain cast two or three weeks. Even in cases where there is that "accommodate bones" (and in this I also had the experience in the flesh) becomes unnecessary the plaster (plaster). Also i had an accident in Nicaragua, with dislocation of the knee and even with several weeks of not sleeping in the pyramid. The reminisce of the effect allowed me to finish my trip two weeks after -even loading bags and walks of major airports in Brazil and other countries - with slight discomfort. The lack of rest and I pyramid "step invoice", remaining some discomfort during almost three months, only in some moments of climbing and sports practices. He explained these things because as a scientist sceptical until where the false sceptical would not get; I could not speak of what I haven't proved my own. A football player or any athlete be unable to attend the next party by a sprain or similar lesion, is today a shame for their doctors, completely uninformed about these advances of real medicine.

9. - In the future: I sincerely believe, although i can't vacate still with the practice, that the pyramids are used for much more than heal. And this is saying a lot, but it i think it is as well. Civilizations that built the great pyramids were undoubtedly of huge demographic composition; they were not scattered tribes or neither the twelve tribes of Israel, nor a clump of minority scientists and isolated, but an entire culture, a whole civilization in the broadest sense of the word. Even if they had what today we call "free technologies" because they were not influenced by a tyranny of the markets, the Great Pyramids are not the fruit of a small group. However they left no corpses or these were too few. The officially found up to now, even in the strata deeply removed by archaeologists, don't pass to be of relatively recent times, or post-diluvia and even so, the oldest cemeteries do not belong to times as old as the that is awarded to the great constructions. Without leave to other topics, it may be said to have been found "giants" (and very giants), although on the internet and certain tracts have been disseminated images adulterated and exaggerated to hide the reality. This is only a theory, but the perceptive reader intuit that this is a use of the pyramids that exceeds the simple healing and preservation of health ... And the things we are discovering that lead us to think in a more transcendent destiny of man, that the simple life, with or without reincarnation. If we stick to the "beliefs" of the ancient Egyptians, we find ourselves with a few "gods" who taught the men to be something like super-men, i.e. to transcend to a natural realm beyond the simple mortal human. In "Sacred Technology of the Pyramids" the information on the new archaeological interpretations expands on these issues, but for now the only fact to cure many diseases (almost all the known as physical ailments) and preserve the health, it is something that has come out of the theory and has practical applications as immediate as indisputable.

Chapter III
BRIEF HISTORY OF THE PYRAMIDOLOGY

In the early decades of the twentieth century, Antoine Bovis, Wilhelm Reich, Karen Drbal and a few researchers more, got check that the pyramidal form has special effects. The Czechoslovak engineer Karen Drbal managed PYRAMID the First Patent of a pyramidal device in 1959, with the No. 91,304 /59, for a small pyramid, capable of bringing back the edge of the razor blades and make them render between twenty and fifty times more than normal.

More than a million people have been able to verify the phenomena from the sales of small pyramids to sharpen razor blades have been written hundreds of books on the pyramid energy... But the Pyramidology is still in its infancy in almost all over the world and this is what is most amazing.

Even more bizarre that the fact that it can regenerate thin edges, retain indefinitely the germination rates of seeds and without going into what applications you have to avoid any substance or rotting cure rheumatism, as we've done thousands of people in several countries and tens of thousands in Cuba, from several years ago.

The pyramid and its qualities is one of the most amazing scientific discoveries of the twentieth century but almost half a century after its first patent, thirty years after the first definitely curing rheumatic and fifteen years after the first therapeutic achievements in Cuba, millions of people continue to suffer perfectly curable diseases with the pyramids, not knowing that there is this alternative. That on the other hand, it is the only one who has achieved success actual therapeutic on the rheumatic diseases and sclera treated.

There are about 167 types of rheumatism (and more about forty modalities of complication, the majority of bacterial aetiology). But the official medicine and the numerous delivered to the interests of the international pharmaceutical market, or are unable to establish the aetiology of the rheumatic or a therapeutic perspective that is not the palliative of the analgesia and the corticosteroids to swollen areas, always partial, insufficient and destructive.

One has to wonder how it can be classified the modalities of rheumatism if one does not have an etiological box even though it is only by theoretical symptoms?, or conversely, one has to wonder how it that having been able to divide the symptoms so efficiently has not been reached a etiologic theory. But as this is an enigma impossible to solve without political considerations, and the economic system in which we live, we will leave it up to this sick still being manufactured to sell drugs and we will work to continue rowing against their flows with the means at our disposal.

One of them, that does not produce collateral damage by chemical attack, nor consumes electrical energy and nor does it have more maintenance costs that any piece of furniture, is the pyramid. But the little walk and by more than we wanted to stay between the four sides of one of them, or the eight beams of a structural pyramid, we find ourselves with a long series of problems that would not leave us move forward in the dissemination of the Pyramidology in general and in the pyramid therapy in particular, if not we assist you the due attention.

Why does a half-century of the first advances, only Cuba applied as official therapy? And that very recent, because on October 12, 2005 informed of Dr. Ulysses Sosa Salinas the formalization, for part of the National Science Council of Natural and Traditional Medicine in Havana, after a process of three fifteen years and countless sacrifices on the part of this doctor eminent and heroic. Finally, after which they were issued numerous technical guarantees of the pyramid in various hospitals, clinics and centres related, the December 07 2005 the agency issued the official statement, as can be checked the certification of next page.

There only displaying some of the more than seven hundred documents issued at various points of the world, of which more than half are from Cuba, on the results on the pyramid effect and the "energy pyramid", they want to or not skeptics, it exists because it demonstrates the fact that there are effects. In this blog we talk more on the question of terms; today we are talking about "energy pyramid" the same way that we talk about "wind energy" or "electrical energy". We do not adhere strictly to the whims of the physical sceptical to carefulness, that want to define the energy in four modes, when don't even know that there is only one and the other three are the effects of the same.

We are already tired to discuss, to make so many checks and analysis on the power of the bacteriostatic pyramids, with what we have a biological cleaning, a natural asepsis in our bed or home, when is pyramidal, that cannot give anything else in the world. Nothing was rotting inside of the pyramids and only the symbiotic bacteria to higher organisms can live without problems and within their ecological margins.

In this issue we are a couple of "mysteries" in terms of root causes that make it to the quantum physics, rather than to the microbiology, but the case is that these amazing discoveries only the we are taking a few tens of thousands of people in the West (many more all over the world), a few tens of thousands in Cuba, while more than a hundred million Europeans and some eighty million Americans suffer from constant headache by rheumatic affections.

Motor vehicles took only in grafting, the same as it took the oil producers to provide the necessary provision. The aircraft took less than eleven years in grafting (in 1903, the Wright Brothers' first flight and in 1914 fought with machine guns synchronized to the propellers). In eleven years, something as complex as a plane was in the two sides of a world war.

CENAMENT
centro nacional de medicina
natural y tradicional

El Consejo Científico del Centro Nacional de Medicina Natural y Tradicional (CENAMENT), considera válidas las evidencias clínicas presentadas sobre el valor terapéutico del EFECTO PIRAMIDAL en una amplia gama de dolencias ortopédicas. Aun cuando se recomienda continuar el desarrollo de investigaciones que permitan ampliar la muestra y aportar nuevas evidencias en Ortopedia y en otras especialidades médicas, el resultado de las investigaciones demuestra su capacidad como:

a- Anti-inflamatorio
b- Analgésico
c- Bacteriostático
d- Miorrelajante
e- Sedante

Por lo que, en uso de las facultades que le son inherentes, este Consejo Científico aprueba por unanimidad el empleo del EFECTO PIRAMIDAL en aquellas patologías donde se justifique como tratamiento de algunos de los síntomas y signos antes mencionados y recomienda generalizar esta terapéutica en el Sistema Nacional de Salud.

Dar a conocer éste resultado a todos los posibles interesados y publicar el presente dictamen en el Sitio WEB de Medicina Natural y Tradicional.

Dado en La Ciudad de La Habana, a los 7 días del mes de Diciembre del 2005.
"Año de la Alternativa Bolivariana para Las Américas"

Dr. Leoncio Padrón Cáceres
Presidente Consejo Científico

Calle 44 # 502 esquina 5ta. Av. Miramar. Playa. Ciudad de La Habana. Cuba. CP 11300

In 1027 Antoine Bovis makes the pyramid first discoveries. A long time after, even the unknown mass these discoveries. Is it because we could not be found application war? Will we find the keys to the mystery in political affairs and of the market rather than in physical matters, medical or any science that has nothing to do with modern politics and anthropology?

The first scientific team Osiris, formed by two physicists and a mathematician (apart from a campus auxiliary, peripheral and separate, composed of biologists, chemists and doctors) had two goals completely unrelated to the therapeutic issue, but the director of the same had healed many furry friends since ten years before and himself a crippling rheumatism. However, our purpose was to discover how and why they occur the pyramid effects, to deduce from this the level of knowledge and way of thinking of the peoples builders of great pyramids in antiquity. That is to say, a purely anthropological purpose. Finally, I finished worshipping quantum physics and they ended up loving the pyramids.

In principle, the only physical thought of them as a simple curiosity mechanical geometry, and the magnetism, which allowed better consider the neutrinos, but then became for all in something that downplayed the enigma motivator of those who and when they built the Great Pyramids.

That all knowledge is deep, all mathematics, geometry, physics or chemistry, reaches its degree transcendent in the lessons of the Pyramid.

In psychology, anthropology and sociology, all perfect structure, all layout really useful and functional, brings us to the shape of a pyramid, both in the abstract as in its material expression. The mathematician of the team arrived to write on the cover of one of the folders: "All forms are important in the universe and that is why there is, but the pyramid is the perfect way and absolute, even on the field and in spite of its importance in the spatial dynamics. The area represents the stillness, the hopes, the neutral and stable, preparing in the smaller volume, the greatest place to life, but the pyramid represents the universal activity, life itself, the evolution and Creation".

On the part of the physical, the issues "pyramid" accessory was secondary. They are only interested in considering the sub particulars and in particular the neutrinosand knew that the pyramidal shape the builds up spontaneously when it is functioning properly. At the end of the investigations, which lasted nearly six years, the primatologist was obsessed by the quantum and the physical by the pyramids, but the margin of these curiosities, what is most amazing remains that any of the discoveries made, such as the real shape of the water molecule, which confirmed a prior work by the two times Nobel Laureate, Linus Pauling, still barely known publicly.

THE MOLECULE OF WATER IS NOT H2O

Or is that the world has not learned that the water is not H_2O, but $[5(H_2O)]$ and even more incredible, is that having been published in a protocol that received at the time all the universities in the world, just have some web pages that reflect the issue and only teach as "anecdote" in the upper courses of some careers...

What would be important or dangerous to teach things like that? Because very little and much, or vice versa, depending on the person concerned. For chemicals that are going to work as employees of a factory or as professors do not have major interest since their calculations did not vary. The monomers of H_2O that make up the polymer water, whose true molecular formula is $[5(H_2O)]$ can be used as a basis for practical purposes and normal conditions.

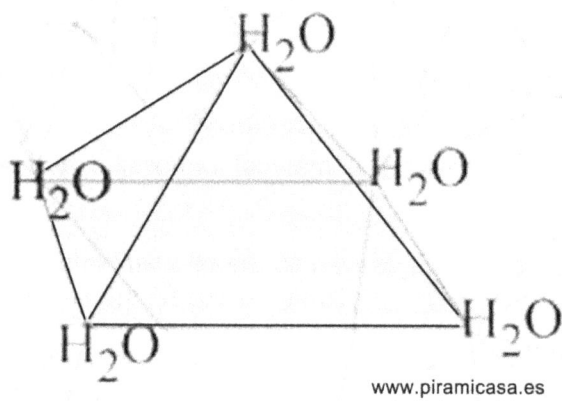

www.piramicasa.es
www.vitalpyramid.com

Treating water in a pyramid, we recover its true form, its perfect symmetry. The molecular simetry have 51° 51' 14" of inclination of the faces.

But for aetiology, a researcher interested in studying the causes of the rheumatism, this is a matter of vital importance. It is necessary to understand the difference between water components of the synovial fluids in well-structured molecules that do coincidentally?

Have perfect shape of a pyramid (or with inclinations of faces of 51°51') and the water malformed, with broken links, where the molecule pyramid breaks down easily and is a ruin, because the monomers loose in time to dissolve and lubricate are converted into oxidizing and deforming the structural DNA and almost all organic molecules. These water molecules, weak in their neither links, nor can traverse normally been semi-permeable membranes, because to be bent to allow their interstices is saturated with other molecules. In such a way that the water, molecularly speaking, becomes a kind of "mud", consisting of the monomers $H2O$, the uric acid and toxins that water should dissolve and drag, rather than contain. Because while failing to recognize that difference, rheumatism will remain "of unknown aetiology "and will continue saying that "there is no known effective cure ". This is not the case for us and we hope that it will not even apply to the dear reader. But as long as we leave other most profitable activities, we sacrifice other personal projects, we risk to receive the peasants used their shovels that gives all struggle, to disseminate the Pyramidology, while we still are wondering: How much do you need to continue to suffer the humanity to awaken the system with which it has been left enslave?. Are you going the Pyramidology to its rightful place in the twenty-first century, or will be - like so many other wonders - something used by a few "modern magicians"?

At least has the Pyramidology, an entire country in which there is no loss, where the rheumatics are increasingly being treated, exist then increasingly less, because there are no rules of financial market, but needs that are filled only with things efficient to one hundred percent. In the rest of the world things are different. We have attempted in this fight, stick to the rules of the market system in which we live, to comprehend that it was the only way to be able to deliver what has been revealed. But our big surprise was the discovery that there are a lot of good people who "accuses us" of traders, speculators, as if perhaps someone could deliver a product free of charge, as if our purpose in life was exclusively earn money, (while it is the only purpose of the majority and especially those who we diagnose) and when in reality we have all been spent for decades to reach the discovered. To disseminate our work is considered SPAM in many lists of internet and media.

The laboratories announced the new drugs by the television as "big news" and little lack to pay them, instead of them. We cannot understand these contradictions; if someone explains this, to change offers you free a lot of information about pyramids. People who are called "sceptical" occupies spaces web, radio, television, etc., and is devoted to stigmatize the Pyramidology when have ever done a minimum experiment to authorize them even to give an opinion of beginners. But none of them are encouraged to make a complaint or a trial for the reasons that say they have against us. While the primatologists (called "pyramidiots" by these false sceptical) we can demonstrate what we say, they only show that the spirit of the Inquisition is still alive, even if use another costume, other pretexts and other means. As the reader will understand the mysteries of the pyramids, the most incredible thing of their effects, are nothing compared to these other enigmas of the ignorant that deny what do not know, the suffering that prefer believe in the saints, the scientific issues unresolved in full "space age", the unknown etiologist, the diseases while foray computed values are manufactured 'nanites' and are photographed sub particles...

The pyramids have not lost the charm of the mystery, because we have revealed, but the barbarism technified that we call "civilization" has mysteries much more difficult to resolve and for nothing charming. But in matters of health, the greatest mystery is the market. Does anyone know the names of those responsible and investors of the pharmaceutical laboratories international? Perhaps they have some answers, but some are worried about because of the advance of the Pyramidology while others are considering investing in this ancient and at the same time revolutionary, ultramodern therapeutic form.

Chapter IV
PYRAMIDS AND ANTIPYRAMIDS
(the interior of the pyramid material)

1) Sweep (excludes) free radicals
2) Accumulates neutrinos.
3) Dehydrates
4) In therapeutic, it is the analgesic and anti-inflammatory since the first treatment session, but its analgesic potency, although smaller than that of the anti pyramid, is notable in almost all cases.
5) Selective is bacteriostatic (prevents playback of saprophytic and parasitic bacteria).
6) Activates all the organic processes sorted and prevents entropic, such as the putrefaction.
7) You can remain in its inside all the time that I want to do. While more, so much the better.
8) Its effect is slow and gradual, depending on the materials. The pain goes away for a bit, but definitely, as the ailment is gradually disappearing.
9) Was form provided that the pyramid meets with your requirements (paramagnetic material, correct orientation, levelling, proportions, etc.).
10) Has No known contraindications, although there are already thousands of users around the world. More than one hundred and thirty only in Spain, in January of 2007.

Lower ANTIPYRAMIDS: Below the plane of the base. Top: Top of the vortex
1) Receives the free radicals excluded from the pyramid and the collects.

2) Does not accumulate neutrinos, but the water molecules there treated after they do.

3) Also dehydrates but less that inside the pyramid.

4) It is anti bacteria too, but a bactericide less selective, so that kills a broad spectrum of bacteria, even some symbiotic. Why cannot be used without further medical knowledge? It expands on the information below.

5) Activates some biological processes and slows down other. It also prevents putrefaction.

6) Can only be done short exposures of 20 minutes up to 50 minutes maximum. They are used almost always small pyramids.

7) Effect more powerful and faster than the pyramid. That was way before the pyramid has visible effects. It is analgesic also.

8) Do not always form. Prevent its formation the irons of the floors and the ground.

9) YES has contraindications and its use is advised by therapists or with medical supervision.

10) We are well aware of the effects of the ant pyramid lower, but the upper ant pyramid, weaker in its composition also known quantum but incomplete, has different behavior, very noticeable, but little investigated, so that we don't advise its therapeutic use at the moment. There is to do more physical research of that aspect, before moving on to the therapeutic trials.

In Cuba is not used in pregnant women, patients with pacemakers, heart disease and severe in alleged cases of individual sensitivity. Of course, refers to the ant pyramids, not the beds or houses, where it is used inside.

It must be borne in mind that the ant pyramid, although doesn't form on the floor because the magnetic currents telluric (0.8 to 1.1 Gauss 0.5 against the power of the earth's field in the air), quantically decompose and drag the ions that would tend to be retrained. But the residual effect (gradual build-up of heavy ions and other atoms) will in that soil, after removal of the pyramid, not grow some plants for some time (depending on the magnetic susceptibility of the species).

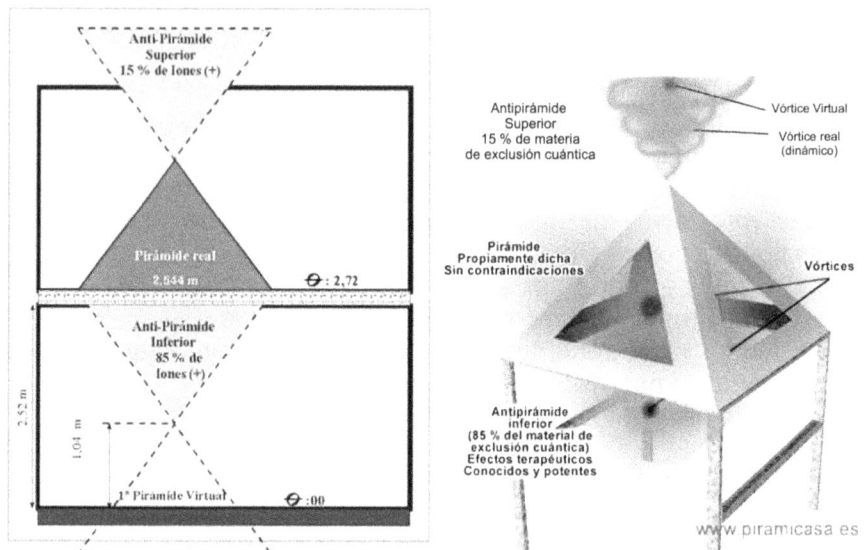

Nor is there a way on the floors of the apartments, especially when there is a little bit of iron in the. Just a few cables of copper or iron passing beneath to avoid almost entirely the formation of the ant pyramid. Let's say to form can do so when the pyramid is hung, on a wooden table or glass or on a wooden floor.

WARNINGS on the UPPER ANTIPYRAMID:

Is that forms on the top corners of the pyramid material. Does not work as well as the lower and although have been partially studied their characteristics, requires further investigation to determine its use without risks, so that it is NOT recommended to use pyramids under the bed or a wheelchair, which is already upper ant pyramid could cause undesirable effects. We know that stimulates the endocrine system, so that positioned and oriented under a chair you can excite the sexual glands, but these practices are not recommended until we know more about the particular.

[Update 2016] There are some therapists and doctors working on this issue. It has achieved extraordinary healing effects, but still have to investigate further. We know (from knowledge of physics) that the top anti-pyramid is more potent than anti-pyramid below. We now know that it can also be used in therapy, even with better results. But we can not yet promote the use in private, home therapy. Therapists need further investigation.

Chapter V
GENERAL GUIDELINES OF THERAPY

We will divide the pyramid in this chapter:
(a) PYRAMID itself, i.e. the interior of the pyramid material;
(b) ANTI PYRAMID, or field projected down from the plane of the base.

PYRAMID:

In the interior of a perfect pyramid there are no contraindications, except that it has a very large density, with more than 200 kg. /m3, which is closed, and with a higher density of 120 kg. /m3 or that its size according to usual material densities, with more than six meters on each side of base. In these cases not have harmful effects, but if very unpleasant because of the rapid loss of free radicals in the mucous membranes of the stomach, that the brain registers without understanding and reacts instinctively, evacuating by all parties as far as possible (diarrhea, cystitis, sudden, vomiting, etc.). This is prevented with a device which slows down the pyramid effect without prevent it, so that the exclusion of radicals is performed slowly and gives time to our body to become accustomed to the pyramidal environment without feeling anything. In a bed or a pyramidal house correctly built we can remain indefinitely and while more time, so much the better.

For healthy persons or those with mild aches, apart from refer definitely the majority of the bumps of health, will be preventing and avoiding a wide spectrum of diseases, but in people affected by rheumatism, sclerosis, fibromyalgia and disorders of the style, as well as acute or chronic infections, loss of the intestinal flora, etc., it is convenient to use pyramids closed, so that the power will increase the effect, reducing the time of treatment. However, a person may continue to use it as usual bed. In some cases it may be preferable to remove the coverage once achieved full recovery, leaving only the structural part as in the normal pyramid house, since it is only will try to continue with the maintenance of health.

The pyramids of intermediate sizes (between eighty and 120 centimetres of basis) serve to combine pyramid and ant pyramid in treatment on members, since there is no body, but if the arms and legs.

The overall effects of the pyramids are varied, depending on the status of the user, but in any case you will notice relief to any pain, inability to get infections, colds, etc... The pyramid doesn't make us "Superman", so it is possible to get a fever caused by constipation, as a reaction to toxins, as in any case of poisoning. It is the only reason for fever found in users of pyramids to sleep.

Currently 13 we have many cases of pregnant women sleeping in pyramids, and there are no contraindications if the pyramid is absolutely paramagnetic, while the case is quite different from the ant pyramid, as explained below. There are some cases of heart disease in Spain that sleep in pyramids, one of them with pacing, and two more in the USA, also with a pacemaker. None has had any qualms and one has presented a remarkable improvement; although for a reason of prudence has not stopped the treatment, you have reduced the medication. In chapter VIII answers the most frequently asked questions.

Pyramid Effects in anti-reheuma therapy and times of aplication

We know that the pyramid is the best medicine for rheumatism. But the antipyramid also has much to contribute to the New Medicine

ANTPYRAMID:

As can be deduced from the explanations Chapter IV, the ant pyramid is convenient, simple and quick to use, since it is only accurate manufacture a small pyramid, of between 30 and 50 cm side of base, closed or only structural, with the indications given in the Chapter VI. But with the contraindications that owns it is desirable that it be a doctor or therapist who make applications, or in your default who guide and monitor the patient.

The therapies are carried out according to the diagnosis and the possibilities are many, but we are going to discuss in this book only some of the most common, by way of example, since in the book "Therapeutic Revolution of the Pyramids" (spanish) is detailed with a wide range of uses and applications made so far. You must maintain a strict control of the time, in order to make the most effective therapy. As a general rule are intervals of two hours between the end of a session and the beginning of the next. More information about times of aplicatioon: http://www.vitalpyramid.com/en/hygiaprospect.htm

The session of exposure to pyramid effect last according to the parties to treat. System osteo-articular and muscular: In the case of the muscular system and bone, not having involvement of other organs, you can extend the ant pyramidal therapy up to a maximum of fifty minutes in house meeting, which is the maximum time given in cycles of flocculation colloidal in the cytoplasm of a cell. More time could be detrimental to the cells. OR maybe with anti pyramid, 50 minutes is the maximum recommended time of treatment.

Internal Organs: According to the pyramid that we use and the severity of the condition, it is possible to extend up to 35 minutes the session, but the average is about 30 minutes to treat intestinal infections and serious stomach. Liver problems have been few, but in almost all has been achieved excellent result. The maximum amount of time recommended for liver ailments is 20 to 25 minutes per session. I have no updated data from Cuba; where over the course of 2006 has been treated to several thousands of patients. In any case, treatments may vary depending on the conditions of the patient, so that I do not tire of repeating that the ant pyramid must be applied with medical control, both to avoid risks and to maximize their benefits.

Respiratory System: It is very much preferable the treatment combination of pyramidal and ant pyramidal. That is to say that the patient to sleep in a pyramid, making four or five daily sessions of ant pyramid, with a maximum time of fifteen minutes from exposure to chronic problems and up to 30 minutes for acute problems such as pneumonia or lung infections that require antibiotic treatment. If this is a sick heart, will be viable only sleeping on the pyramid, but it is not recommended the anti- pyramid.

Endocrine System: special care must be taken with this part, because the glands are very sensitive to the changes and the magnetic ant pyramid is a modality of magneto therapy, although we use the soft magnetic field of the Earth. In some glands such as the adrenals seems to have a good resistance to the action anti pyramidal, but in the thyroid some people feel discomfort after ten minutes of treatment, so it is recommended that you do not move from that time in the session of application on the head, because the glands that we are above the neck, except the lymph nodes, are both or more sensitive than the thyroid.

The lymphatic system has a resistance to magneto therapy something superior to the rest of the body, but only a doctor or a trained therapist can determine which is the right time, that in no case exceeds fifty minutes. In France, we have a very strange case Delphine Paillard, a 28-year-old woman with deficiencies in the lymphatic system and a chronic leukaemia diagnosed in 1995, who has studied medicine as a challenge to his situation supposedly lifetime. His physical appearance was a real disaster caused by the drug experienced in it and the sedentary lifestyle.

A year after starting to treat it with a colleague anti pyramids its situation was optimal and now it seems another person, Delphine completely liberated even of the side effects of the medication. Some of the cases treated in Cuba are truly impressive, but this is broad in another book.

RHEMATISM HAS CURE, but the rheuma is not a localized problem, but is a molecular disarray then is need treat whole body. Also occurs with multiple sclerosis and others degenerative diseases. You can not cure these ailments with anti-pyramid, but only can get a relief. To cure these diseases affecting the whole body, is need a standard Pyramid-bed in mild cases, or when a severe case is need a Hercules model.

Chapter VI
HOW TO MAKE and USE pyramids...

To make homemade pyramids for various uses, attached is a template for 50 cm base. The tabs on the sides serve to paste it on a wood. The aluminium of 1 mm is very hard even for a good scissors to cut off cans, so that if there is no shear for metals, it is preferable to do cardboard, farandole - inside and out-, with thick aluminium that is used in the kitchen. Its usefulness, though less powerful, it is virtually the same as that of aluminium plate of a millimetre, but much cheaper and easy to make.

MATERIALS:

NEVER use ferromagnetic materials or diamagnetic," In particular, never use pyramids of copper to treat anything. Its possible usefulness in the future is related to extraction of electrical energy from systems that take advantage of the magnetic field earth phenomena, but the pyramids of copper are truly dangerous to living organisms. In Russia are used to decontaminate groundwater deposits. You should only use paramagnetic materials, with which there is no risk in the modalities advised.

Aluminum is the only metal advised, but only one found in alloy with silica, not with iron. The manufacturers that we use are, in addition, high purity, with the minimum of iron achieved in industrial laboratory, so that the ferric content is negligible.

Other materials can be glass (the best, if it were not for its fragility), wood and derivatives, natural resins with fiberglass and in last case plastics. These do not serve to make pyramids for sleep. It must be borne in mind that these small pyramids have some utilities limited, that should not be used because its anti pyramid top we have not even knowledge for managing this factor, but we can use the lower anti pyramid (and is the most used in Cuba, in hospitals and health centers various), with the requirements and precautions that are listed in Chapter V

PROPORTIONS:

The pyramidal proportions themselves are those of the Great Pyramid of Giza. It is not the world's largest pyramid, since the Bosnian, like the Great Pyramid White of China around 500 meters of base, but the Egyptian is the most perfect ever built.

Its proportions, which give an inclination of 51° 51' 14", are exactly those of the water perfectly structured pentapolimeral in its way. As we move away from this proportion the effect is less powerful and less harmonious. In some pyramids of equal or greater height as the side of base, there are powerful effects but not as harmonics.

There are some very exact formulas but complicated application, as well as to small pyramids we use this formula extremely important.

HEIGHT: Base /1.570845
BASE: Heigt x 1.570845
RIDGE: Height x 1.4946
APOTHEM: Height x 1.269

RATIO MASS: The mass of a pyramid determines its density. That is to say weight on volume. A pyramid of edges and bases very fine won't reach to close its field and will not work. You must have a density related to the percentage covered faces, apart from some weight ratio. The pyramid closed, on the other hand, operates with a minimum of density, but equally its strength varies in function of this. It is not the same a closed cardboard pyramid that closed a pyramid of aluminum. The pyramid beds weigh, depending on their size (only the aluminum part) between 23 and 40 Kg. When they are closed (model Hercules) therapy for more severe, you can weigh a total of 200 Kg., according to the design and apart from the wooden structure that serves as a tatami.

You can use this scale in inches, centimeters or millimeters, because it is proportions.

PROPORCIONES DE LAS PIRÁMIDES PERFECTAS

ALTURA	BASE	ARISTA	APOTEMA	ALTURA	BASE	ARISTA	APOTEMA
100	157,0	149,4	127,9	600	942,4	896,7	767,4
150	235,6	224,2	191,8	650	1021,0	971,4	831,3
200	314,1	298,9	255,8	700	1099,5	1046,2	895,3
250	392,7	373,8	319,7	750	1178,1	1120,9	959,2
300	471,2	448,3	383,7	800	1256,6	1195,6	1023,2
350	549,7	523,1	447,6	850	1335,1	1270,4	1087,1
400	628,3	597,8	511,6	900	1413,7	1345,1	1151,1
450	706,8	672,5	575,5	950	1492,2	1419,8	1215,0
500	785,4	747,3	639,5	1000	1570,8	1494,6	1279,0
550	863,9	822,0	703,4	1050	1649,3	1569,3	1342,9

www.piramicasa.es - www.piramicasa.eu - www.vitalpyramid.com
Tlf: (+34) 639 28 47 87   piramicasa@gmail.com

Some people built pyramids with thin rods, but using stainless steel. This is cheaper, also have effect and with less density, but they are certainly dangerous, because as already explained, the ferromagnetic are harmful. I have had several inquiries from people that have been used, worried about the symptoms after a few days of use (headaches, dizziness and nausea). These are very unpleasant but soon passed through bathrooms with tincture of iodine in the bathtub. ONLY USE to sleep pyramids of aluminium of high purity, wood (and derivatives) or glass. Iron is harmful, and the copper worse. The pyramids of resin and fibreglass are also good, but require that a high percentage of the faces are closed, so they are less comfortable and it is difficult to find natural resins to allow for a good job with the fibreglass. The synthetic are not suitable because they are very slight, although these can be used in the paint on the aluminium or in the decoration of the tatami. Plastics, in addition to generating some static and have a unpleasant vibration and away from the 440 Hz, have some toxicity and would require a large mass to achieve the effect, so that only serve -some- in the paint. Why are not recommended in pyramids closed or open, large or small.

ORIENTATION AND OTHER FACTORS MAGNETICS:
Must have a face exactly orientated to the North (magnetic, that is to say that there is to use a compass). With a diversion of five degrees effects are very scarce and eight degrees usually stop working, depending on the site of the world in which they were found. Some operate with up to 15or diversion, but it would take a very long time explaining these oddities. It must be borne in mind that a pyramid disoriented is like a car without fuel, by good it is.

The pyramids are hung, to use the anti pyramid, must be tied with wires from two vertices to a wall or other fixed component, so that is not disorient.

The small pyramids can fail if they are on a point normal geopathic (small). The possibility that fail a pyramid of 50 cm side is approx. one at 282. If that happens you can simply run the pyramid one meter in any direction, to take it out of geopathic area, in addition to take into account this point not to remain on them. A pyramid bed well done is not affected by these points, but moves the telluric currents. It would not work must be above a geopathy very powerful, of which there are in rare sites or in buildings under which it has strong currents of water systems or ditches of mills. Equal is usually sufficient move a couple of meters the pyramid or if the bedroom is very small, a room change. Any pyramid -with more reason small - should be as far away as possible from metal masses such as refrigerators, engines, grates, etc. . . The prudent distance is equal to twice across the base. In the pyramid beds there is greater mass, therefore the tolerance is greater, so is usually sufficient a metro of margin.

LEVELING: While best level is, the better work. Works just like an inverted pyramid, but with some interesting variations and not recommended for application on living organisms to have better designs, for which more experimentation is needed.

Do not have practical applications by now. There is a model developed in Holland some years ago, consisting of an inverted pyramid within a larger pyramid; but the only benefit it has been the neutralization of some geopathies, think that equal is achieved in a normal pyramid. Pyramid a normal, well-built and well-oriented, is your thing and already offers a huge field of utilities.

USES:
One pyramid of 50 cm of base is used to treat minor injuries, wounds

rheumatism of wrist, hands, or feet, but they have to follow the instructions of the doctors and a strict discipline of application: In cases of rheumatism and wounds, on 20-minute sessions with TWO HOURS of interval between the end of a session and the beginning of the next. But bear in mind that the treatment may be necessary for several days. It is also important to understand that a small pyramid is not the same as sleep inside a pyramid. A segment of ANTI-PYRAMID (that is the mode of use of this template), you cannot even begin to compare with spend eight hours every day within a pyramid bed. REMEMBER: The rheumatism need a complet treat of body for eliminate cause definitely.

OTHER USES of this pyramid of 50 cm: (The door is not necessary)

Treat natural medicines from any source, except those containing copper. Nor should they be treated psychotropic drugs and in general no allopathic treat synthetic products. If you can deal with allopathic minerals, such as virtually all salts. This may be water, food in general, oils and creams to use external or internal. Floral essences, homeopathic medicines and everything you need to retain the so-called "subtle energies", should be treated a maximum of four hours for a security issue. Equal could be dealt with in days, but four hours later (provided that the pyramid is loaded in a few days before at least), is when the medications obtained the highest power and stability. If you leave or homeopathic flower essences more than three days, they will be better molecularly structured, but not always more powerful.

Just as it is possible that lasts for more its quality, because conservation is indefinite chemical inside the pyramid, but don't know how much would last the subtle properties. The homeopath experiments indicate a loss of dynamic (determined by statistics in application) after four days, but have not followed the experimentation by most days to see if it is retrieved and/or stabilize, as occurs with colloids, on the grounds that patients should be treated with security. Those who work with floral essences are with the same dilemma, but it would be necessary to experiment more.

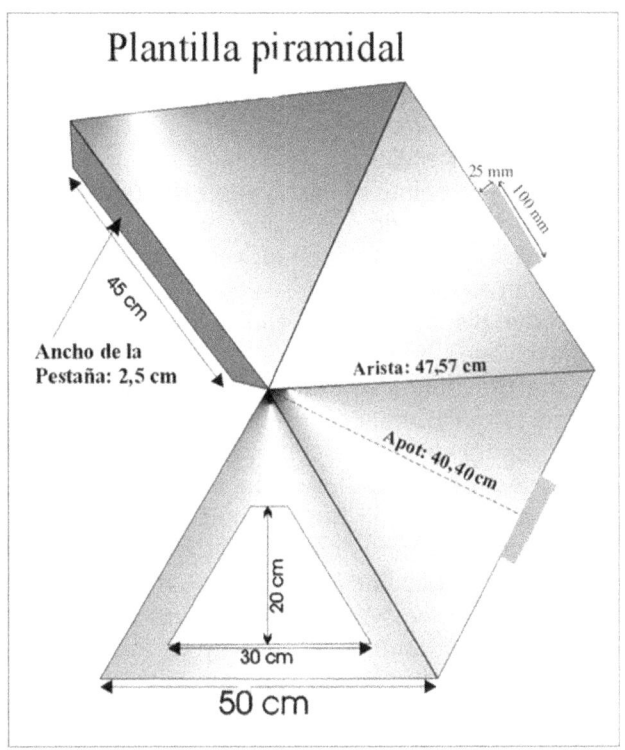

Perhaps the ideal would be to deal with the medications several days prior to revitalise them, as well as the recipients before adding the essences (floral or any other).

Chapter VII
PYRAMIDS, EFFECTS, AND ITS USELFULNESS

In this section we will refer almost exclusively to the pyramids for sleep.

1) What are they for?

    2) Who can use them, aren't uncomfortable? Serve to...?

    3) Do they have contraindications? ...What pregnant women?

    4) Can you have excess energy?

    5) What if they form anti-pyramid, which passes with the neighbours from below?

    6) What are its paranormal effects and uses spiritual?

7) How many people sleep or currently living in pyramids?
8) What fits well in my bedroom?
9) Why are not sold in all stores?
10) What decoration is going?

1) WHAT ARE THE PYRAMIDS?

There are four basic points of utility, apart from the use of any bed:

a) Therapeutic: Is the primary utility, both for power: Antioxidant, relaxative (muscle-relaxant), antibacterial, anti-inflammatory, anti-rheumatic and sedative.

The rheumatic diseases of any etiology lead the field in the results and although have been thousands of cases treated, all have been successful. But the prevention of many other bacterial diseases, stones and all the related to malformations of the organic liquids, are important palliative or are completely overcome, depending on the patient's conditions and the etiology of the condition. Increases the immunoprotective function in all human bodies, animals and plants. It regulates the endocrine system; there are experiences from important corrections in immunodeficiency problems until disappearance of diseases related to the same. Wounds heal more quickly and his power bacteriostatic reduces or cancels according to the cases, the risk of infection. But what you accomplished in general the user, who is sleeping or living in a pyramid, is to remain healthy, vital, relaxed and balanced energy. Other biological effects - explained in the other books - contribute to enhance and prolong cell life and therefore, the life of any person, animal or plant increase in its size and quality for some species, and with an important resistance to attacks from pests. The germination rates of seeds are maintained almost indefinite in the pyramid and endure over time after being removed from it. Nothing decomposes within a pyramid perfect well-built and installed, although the orderly processes of digestion (intestines animals) are not interfered. % B) Bio-regulator and conservative: The seeds of any plant, as well as their seedlings and seedlings will be reinforced vitally, reaching 30

c) Analgesic, anti-inflammatory and other properties and also in the use of ANTIPIRÁMIDE (the field located below the plane of the base), but this usage pattern corresponds only to therapists with experience, because this is a similar effect to the magneto therapy. In the pyramid beds is not used the anti pyramid effect, but that it remains on the inside and never below

d) Energizing, relaxing and psycho-regulator: These effects are subjective aspects that only are explained as the person is feeling the effects. Talk about it before, is to induce theoretical speculations, pseudo esotericism mystical and false expectations. There is no "magic" outside of nature, because nothing escapes to the natural laws. The psychic effects are normal of a brain that works physiologically well and able to expand and develop their potential.

2) ANY PERSON CAN USE?

Any person who does not have to panic the "strange things" and to new technologies, you can sleep or live within a pyramid. Fear and ignorance are the reasons that many people still have reservations about it. Suffering from diseases that could be avoided and prefer to think in "the little pill" or the saints ... The idea crushed till boredom academic by the dictatorship, that were "large tombs" Egyptian, Mayan, etc., no longer convincing to people thinking and cultured, because this understanding that never were tombs (nor made by these civilizations already retrograde) and the amount of people that are "changing the chip" increases exponentially. We use microwave ovens, mobile phones, computers and televisions, without really thinking about whether or not we can be able to use those things that if they are dangerous and demonstrably harmful to health. The pyramids do not have any of these risks and even attenuated the effects of some of the appliances that we use. Many of the problems caused by these sources of non-ionizing magnetic fields are greyed out or neutralized in their effects to those people who sleep in pyramids, to occur a detoxification of radiation.

There is no age limit for the use of pyramids well-manufactured. There is no risk to children, adults or old people. No disease represents risk to those who sleep in pyramids, but quite the opposite. They can use them the healthy people to preserve the health and the sick to assist in any therapeutic process. With regard to the comfort, not there are complaints.

There have been no reported feelings of being overwhelmed or claustrophobia and the feeling of the majority of users is to feel protected. Even though the sensation is subjective, the cause is real, because there is a protective effect against true geopathies and mild radiation, as well as against internal organic factors (such as the bacteriostatic action that prevents the reproduction of saprophytic bacteria and/or interference)

Many of the questions received have been referred to if you can practice well sex in a pyramid bed. Seems to me to be a trivial question, but deserves a response: the structure, as well as tatami mats, it is very resistant and strong, so you can make multiple "acrobatics". Other people assume that can replace the Viagra and other drugs by the style, but although the pyramid does not have these effects in live mode (or their contraindications), which makes is to restore the balance and overall organic inducing some cathartic processes in terms of psychological health, with what many problems of sexual relationship can have a indirect solution.

3) DO YOU HAVE CONTRAINDICATIONS IN PYRAMID USES?

Only one: the intake of alcohol or barbiturates or psychotropic medications two hours before entering the pyramid because it enhances the effects. If you've eaten too much, it is preferable to put the head to the South (valid even for those who don't sleep in a pyramid). But if you have drunk alcohol, it is better to wait on the sofa that this has been metabolized. Nor is it recommended that a person between with fears in a pyramid, in the same way that it is not advisable to do so in an elevator if you have claustrophobia. There will be no physical or psychologic damage, but if a magnification negative (purely psychological) of sensations. Pregnant women have no downside but for them as well as for the babies, the effects are absolutely benign. A amniotic fluid molecularly improved in its composition of water is the perfect environment for the development of the child. Pregnant women who have entered in the Great Pyramids of Giza, whose power still endures both or more that a pyramid bed, have not had any problem. To the date of this update, there are many women who have passed their pregnancy without problems and with notable benefits for themselves and their babies, during and after childbirth.

4) CAN YOU HAVE "overload" OF ENERGY?

Only when it comes to pyramids built with inappropriate material, ferromagnetic, diamagnetic," or with an excessive percentage of these, such as copper, zinc, etc.… In pyramids of hardened aluminium to silica or wood, although it is reach very high rates of energy, the quality of this wobble will determine excellent quality of the effects. A pyramid was built and installed correctly can cause strange sensations in the beginning, to a person extremely sensitive, when the body is experiencing for the first time, but such rarity will be the product of a psychological interpretation and/or brain conditional, that changed very quickly to assume the body benefits you receive.

In hours or days at most, a person becomes used to the pyramidal action and leaves to feel the immediate effects. However, the long-term effects (cellular longevity, endocrine balance, strength of the immune system, regulating blood pressure, etc.) will persist.

These women of the nex photography have over ninety years. They enjoy each of a Pyramid bed. They have excellent health.

5) IF ANTI-PYRAMID FORM, WHAT HAPPENS WITH THE NEIGHBORS BELOW?

In the normal story it doesn't form because the structure of the buildings carried plenty of iron, which prevents the accumulation necessary magnetic to form the anti pyramid. This material leads to land heavy ions ejected by the pyramid and disperses the field anti-pyramid without effects to anyone.

There is only that a careful study of the location to prevent the formation of anti pyramid in the wooden houses, but even so, the most common measures of the pyramid house are 1.33 meters in height, more tatami mats (about 163 cm in total), so that the anti pyramid is often not habitable to the height of the lower floors, but a bit more up. In addition, in the wooden houses is placed a device that prevents the formation of anti pyramid and can be removed to be used therapeutically if necessary.

6) WHAT ARE ITS PARANORMAL EFFECT AND SPIRITUAL USES?

The pyramid is not a "paranormal element", but a appliance, an artifact with physical effect and geobiologic interaction.

Their effects are caused by a set of concatenated physical phenomena and simultaneous, such as those that may occur with an antenna what we would consider "paranormal" to an aluminum structure that captures a hertz lan wave and transmits it to the TV.

Some paranormal experiences (in people prone to them) take place in the pyramid only by the fact that the brain is in best conditions biological, physiological and functional to express their natural potential.

A student simply you will find greater capacity for concentration and retention of what he studies. Someone who works very hard, or a housewife - who works generally more than the male, acting as a cook, laundress, decorator, guiles-plumber cross, educator and several offices more every day - you will find in a pyramid bed a rest much greater than in a normal bed. Your dreams will become more clear and easy to remember in the last few hours of rest, because the muscle relaxation will be more profound and remarkable mental clarity. But this increased "dreamlike awareness" is simply a product of the best organic functionality, not mysterious paranormal properties. People who practice meditation, induction studies conducted using psychic or any practice; it must do so in the last few hours of use, or to wake up. If they were trying to do at bedtime, the more likely it is that the relaxing effects smashing them before you begin the practice. Remember that those who suffer sleep disorders usually sleep like logs from the first night, so the do not have these problems, with more reason fall asleep quickly and deeply. Mental practices, studies, etc., after a break of eight hours in the pyramid, will be not only with a body well relaxed, but also with a brain awake and clearer mind.

7) HOW MANY PEOPLE SLEEP IN PYRAMIDS CURRENTLY?

More people who dare to confess it. PIRAMICASA cannot give data without express consent of the clients, fully respecting the privacy of the same. To contact these people, it can only be done through the mailing list, piramicasa@googlegroups.com . There are forty families living in houses pyramid (not counting those that have rental for life in the Luxor hotels of U.S.A.) In China there are many more, because there are villages with houses formed pyramidal ingenious construction and perfect guidance. In the web of pyramid house (www.piramicasa.es and other sites of Piramicasa). In Spain there is - at the beginning of 2016 - a few 3112 people who sleep on beds pyramid and the number continues to grow.

8) WHAT FITS WELL IN MY BEDROOM?

It is very rare to find a bedroom so small and badly directed in that does not fit a pyramid bed. Even in the case of a diversion from the bedroom to 45° with respect to the north, a pyramid of 2, 10 m side and leaves the step just if the room has at least 3.2 x 3.2 m... The "disorientation" with respect to the walls is no discomfort, but quite the opposite. The body feels much better with the cardinal orientation correct,

regardless of the angle in relation to the walls. We have done smaller steps and more large, or is from 1.90 to 2.30 for marriage and up to 1.50 aside for children. The results with the appropriate materials are virtually the same in all of these measures. See about orientation to end of this list of details.
9) WHY NOT SOLD IN ALL STORES?
Things happen, but it is a product of recent creation, which is being attacked by the alleged "sceptical" that sleep their consciences with theories without making any experiment. Soon will be our pyramids in the specialty shops at rest. Our current prices seem high, especially because it still costs correctly assess the difficulties of manufacture, the enormous costs of four decades of research, the benefits provided to them, nor is it easy to understand that it is not a product "cheap". The angles of aluminium moulds' have required special treatment because are out of all the standards of the market. Now we would like to live in a society of solidarity, non-commercial, where governments to facilitate and manage according to another scale of values, without having to practice the purchase and sale ... As the societies that were able to construct the Great Pyramids, while our economy (apart from the technical limitations), would make it impossible to try a replica of those works. The installation is also something that requires an appropriate learning.

Build a pyramid misguided not pose any risk, but if it would be a waste of money. When we install a basic geobiologic prospecting to ensure the customer that the pyramid is going to work and meet their expectations, but there are very few people with sufficient knowledge. It is hoped that soon there are more people doing courses of geobiology, so very recently begun to count on reliable staff to expand your horizons in terms of sales and installation.
10) WHAT WILL FINISH THEM?
In reality it is a very archetypal element, as old and at the same time as modern, that the pyramid house is decorative in itself and one can only adjust the issue of colour (you can even paint the aluminium by simulating the colour of the wood). The tatami furniture (a measure and assembled with the pyramidal structure), as well as aluminium, can be painted to customer choice, in the range of colours available in the market. We use acrylic enamels ecological that do not require hazardous solvents.

ABOUT ORIENTATION OF PYRAMIDS

The first thing to do is make sure the Piramicama ® (*Pyramidbed*®) (necessarily square base) fits in the bedroom, so we need to know the length, width and orientation relative to magnetic north. Furthermore, we must bear in mind that you can sleep head north (optimal), the South (good) or east (good), but should never sleep with your head to the West. A Normal Piramicama fit anywhere well oriented, but assuming the worst case (45° deflection of the room, relation to magnetic north) is enough space of 3.2 x 3.2 meters. For a model Hercules is the same. In an optical orientation occupy 2.4 x 2.4 meters, but in the worst orientation must have 3.5 x 3.5 meters of space. Explained below.

But first, it should do a little prospecting geobiológica function that can make anyone without knowledge of Geobiology, used for only a compass.

Is to ensure that no important magnetic anomalies. If the compass north mark on all points of the room (logically forming parallel lines of reading), then there is no problem whatsoever. You can go "walking" around the room as indicated by arrows or in any order, but methodically, doing readings at least nine points. Most experts in the use of the compass and with good calculation "ojímetro" do not need anything more, but novices can use a chalk mark on the floor, or make a sketch if the floor is delicate.

HOW TO USE THE COMPASS: always on hand, not on the floor or on furniture that may have screws, nails or other metal parts. It is held at the waist, chest or near the face, but avoid bringing buckles, necklaces or other metals. Few feet away from the walls and the floor.

If instead, as in this other picture-we have a mess where the compass readings different brand us North at each site, magnetic anomalies have always considered "geopathic" meaning "place on earth to illness." In 170 cases (until June 2009) we found only three such geopathies. Two of them were resolved by installing the Piramicama in another room, because it is generally impossible to work under these conditions a pyramid. In the other case the house was all geopathies affected by intense and there could not be mounted the pyramid, must abandon its occupants, since discovering this problem, we learned the history of the place:

All the people who have lived on that site has sickened and even mad in a few months. So (NOT TO LIVE) or at least not have to sleep in places where a compass gives anomalous readings. Most were in nearby groundwater, to building metal structures (that reinforce natural geopathies, etc.), Or combination of both. In other cases, the problem is combined with knots Networks Hartmann and Curry, but these, if not very intense, not usually affect the function of the pyramid.

There can be some exceptions, for example, when the compass is altered only at a point of the room, near a wall or a beam or a metal mass as a radiant heater, in which case we will see if this anomaly is only at that point . In this case the pyramid can work in the room, while no place on this point, and while this anomaly is very small .. Once confident that we can install it without risk geopathies with intense magnetic anomaly included, it is found that the pyramid and determine the best size and accordingly design and user measures (height and weight).

This is OK condition

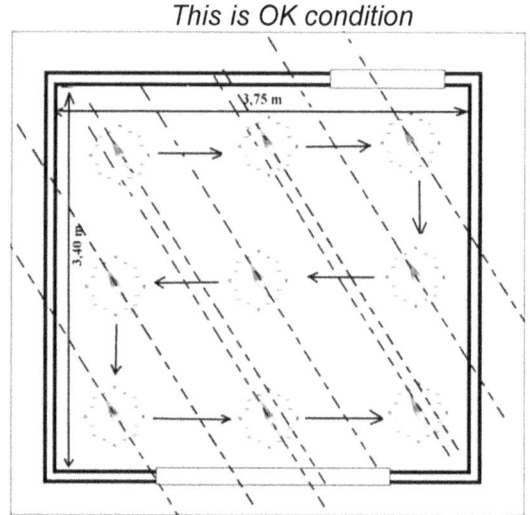

Some examples of bedrooms, its measures and guidance:

It is very rare to find a bedroom that is one of its walls, especially the head of the bed, giving just north. But while Piramicama fit, no problems of aesthetics. On the contrary, thanks to the health body when sleeping properly oriented, irrespective of the distribution of the environment, but also the aesthetics is innovative and often changing aesthetics user has released the mentality of "square", with benefit psychological added.

This drawing is of a typical orientation in apartment buildings, but there are many houses where not taken into account the guidance in building, or city planners laid out the streets regardless of this important factor. To install a standard pyramid of 2100 mm square (2.1 m), is needed in the worst case (oriented at 45 ° to the north), the room has a minimum space available 3.2 x 3.2 meters.

This is the worst geobiological condition (does not work the pyramid, but you can not live healthy at that site.

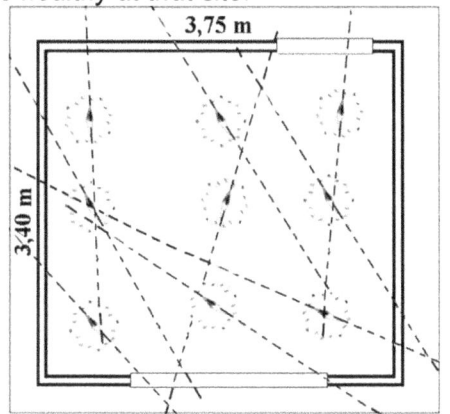

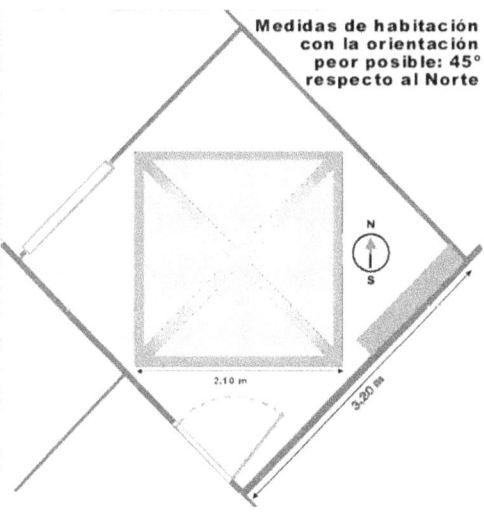

Medidas de habitación con la orientación peor posible: 45° respecto al Norte

In another example we have a fit. The customer wanted a pyramid somewhat larger than the standard, but given the direction, it was impossible. Once assured that there geopatía in the room and asked the stature of users, is that it measures only 1.63 and 1.55 it, so you do not really need a standard size and the extent of 1915 mm was sufficient . The PIRAMICAMA side must always be greater than 25 cm height of the user, but in all cases we try to be a little more the difference, for comfort.

In another exemplary case, a 2.1-meter pyramid does not fit, but if the user's height is less than 1.75 can be installed one of two meters. Without these simple orientation conditions, we cannot pretend that a good pyramid operation.

We have had customers request Piramicama did not want their effects, but simple aesthetics, wishing that we install where it can not work. We refuse outright to install a Piramicama where it can not function properly. In any case, do not hesitate to contact us with any questions, to piramicasa@gmail.com

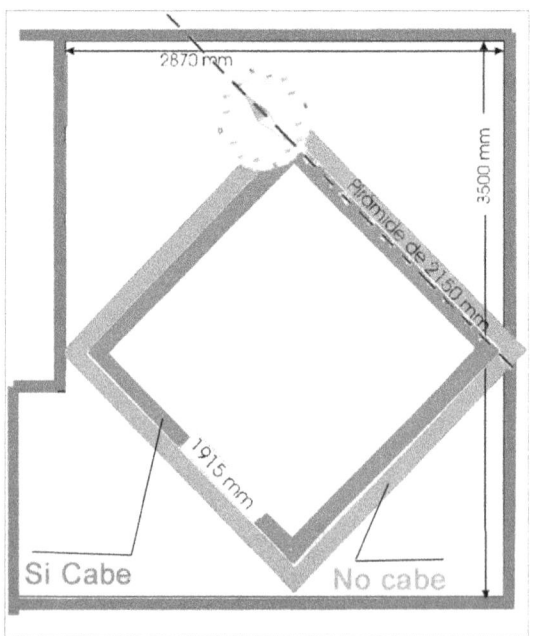

Chapter VIII
SCIENTIFIC BASIS OF THE PYRAMID EFFECT

Without going into matters more long as the treaties in the books mentioned above and others in next edition; let's look at some key points of how and why it produces the pyramid effect.

The discussions of assumptions skeptics who do not do their own experiments is something that we have learned to dismiss, because there is no worse blind that he who does not want to see and any discussion is pointless when there are already so many thousands of convinced from my own experience, but to physicists, who are generally the most practical and interested in go to experimentation, as well as for the general public, are some explanations. You cannot give an informative walk, enriched with images on the page and more information about pyramids:

http://www.piramicasa.es/CONFERENCIA/confer.htm but this is text, a very simplified way, the mechanism of the pyramid effect.

1) Training of a magnetic field of movement faster than normal (in the funnel effect magnetic order). The particles included in this field are accelerated, so that they produce a sort of sweep, breaking down both at the molecular level as any atomic structure not linked correctly, then exclude it from the field or -minority - modify its behaviour within the same, once broken down the matter until the level sub-quantic.

2) Centrifuge inside, which generates negative voltage at the centre (quantum decompression or negative voltage). It is not a "quantum vacuum, but the vortex tends to this entity and sept infarct occurs in a fraction of space which is practically impossible to determine, even with the media with research was carried out the second half of the '80. Today there are probably 24 means similar or identical to determine these phenomena, but it is difficult to obtain access to the groups of elite scientists. The tension-activation produced by the negativization of forces, produces molecular restructuring, at the same time that the breakdown and elimination of free radicals and other harmful molecules organically.

3) Accumulation of neutrinos in proportion to the decompression quantum achieved. This translates into greater vitality for completeness quantum of atoms.

4) SIMAFO (Magnetic phenomenon of sympathy of the Form). Molecular restructuring of the water.

INVOLVES:

A) repetition of the pyramid effect at the molecular level, since the true molecular structure of water [5 (H_2O)] begins to operate as the pyramid in its perfect level, each time that a molecule is oriented with respect to the Earth's magnetic field. This occurs in a tiny percentage of the molecular mass, but enough to continue the pyramid effect at the molecular level, with the consequent raising of neutrinos.

B) Increased tension activity general, but especially of the water.

C) Higher level of solvency.

D) Less or zero oxidant capacity.

THE GREAT "SECRET" OF THE PYRAMIDAL WATER

The true PYRAMID water molecule is [5 (H_2O)]. A single H_2O is a monomer oxidizer, reactive, rather than solvent. This molecule has real shape of a pyramid, with 51° 51' 14" of tilt in their faces ... Exactly the tilt of the Great Pyramid of Gizhe.

One of the main effects of the pyramid well-built, with appropriate materials and working in relation to the planet's magnetic field, is the so-called "Magnetic sympathy of the Form", which in the case of the pyramid is the restructuring of these molecules, thus increasing the end its tension activity, making it the perfect solvent that is in essence, eliminating the monomers "loose", preventing them from forming free radicals and other oxidizing agents. This "corrected water" is the cause of all the cures in the rheumatic diseases, post that the aetiology and deep physical of them is precisely the disorganization of the molecular organic liquids. In the books recommended below will address these issues in greater depth, but it is suggested that the practical experience, since reviewing and understanding the whole theory and analysis carried out by the author and dozens of scientists over the decades, would be for many an unfortunate loss of time. The best thing to do is check the theory while enjoying the practice and benefits of the pyramid.

HOW IT WORKS THE PYRAMID

The pyramid works like water molecules. That is $H_{10}O_5$ which is the same (5 [H_2O]). The Pyramid Effect is the most oldest and natural that exists in relation to life. The water molecule has many missions, but is not only a perfect fluid for the biologic mechanism of the plants, animals and humans, or only the perfect solvent for the homeostatic functions.

The water is vital because have the natural mission of capture the neutrinos, and they are the basic brick of the atomic universe. Our atoms have many "holes", because diverse radiations. The neutrinos are included in the atomic structures completing these atoms. The Pyramid Effect can be compared with pyramidal funnel. But instead of liquids, we talk about quantum particles.

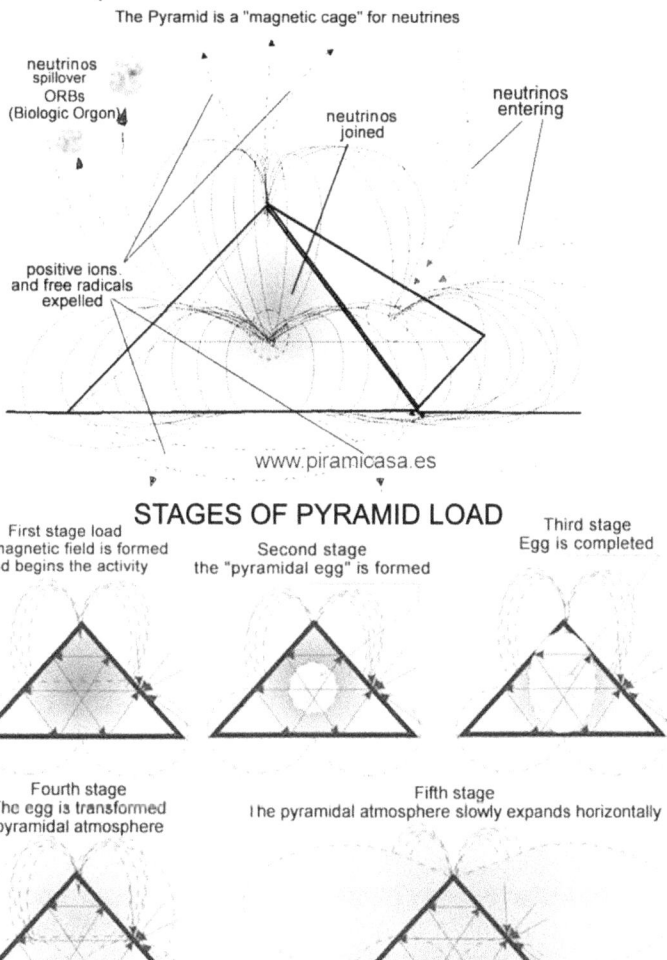

Chapter IX
EXPERIMENTS WITH PYRAMIDS

Do basic experiments with pyramids is something that can make up to small children, but there are more complicated, that may be made according to the knowledge and materials available. Then we divided in stages in school experiments, suitable for various ages and levels. Can also be manufacture pyramids structural and demonstrate that not only serve the closed, but is more complicated than cut and bake a cardboard.

PRIMARY SCHOOL:

First experiment: Check the lower oxidation. Build a pyramid of cardboard or cardstock with the measures of the template or scale of the same. Line it with aluminium foil on the inside and outside for greater effectiveness. Once oriented through a compass, with a face exactly to the north, you will no longer charge for one hour. Then cut an apple in two and a half is placed in the pyramid, a third of the height, in the geometric centre (on a piece of wood or a box of cardboard). The other half is left within another geometric shape, like a shoe box, a meter away from the pyramid. The conditions will do the same? After four hours you will notice the difference.

Second Experiment: Revitalisation of seeds. Perform a few germinate as the usual in several grades of primary school, but try for a few weeks, some seeds in the pyramid and check the difference of the subsequent growth in germination, with the other untreated.

Third Experiment: Revitalisation of the plants. Also with germinations, maintain a in the pyramid and the other outside. What will grow best?

Fourth Experiment: Testing does not rot. Place a piece or slice of meat inside of the pyramid and other equal outside it, in a shoe box. The shoebox will be throwing rotten smell the next day. What will happen with the pyramid? What if leaving more days? Because to experience.

HIGH SCHOOL:

You can do the same experiments than in the primary, with some more complicated or not recommended for small children.

First Experiment: crystalline restructuring of the steel. Some young people already shave and spend a disposable shaver each week. How long does it last if stored in a pyramid? For this is the same as if it is closed or structural and may be smaller, of about 300 mm side, 285 mm of edge, 244 mm apothem, which will be 191 mm in height.

Second Experiment: impossibility of putrefaction and effect in the saprophytes. Place the pyramid in a sample of meat in principle of decay, with larvae of flies newly hatched. You will appreciate a sudden exodus toward any place outside of the pyramid. Not having rotting saprophytic worms cannot be fed.

Third Experiment: in agricultural schools where classes are taught beekeeping can be tested its action on the hives affected by various diseases, treated up to the fearsome disease (varroa, chalkbrood, etc.) In the normal schools it is possible to get some samples of insects not saprophytes and saprophytes, so that you can see the difference in behaviour with respect to the pyramid.

Other experiments: biology teachers can design countless experiments according to the available materials in the school. The agricultural laboratories tend to have more complete for biological experiences, but the techniques and skills tend to have better items for physical measurements. To them, such as the university courses will be served following the saga of experiments.

MORE ADVANCED EXPERIMENTS WITH PYRAMIDS

The anti-aging effect is not a mystery. The cure of diseases as rheumatism, multiple sclerosis, etc., is not illusory, is not a placebo effect or depends on faith. They are well studied and explainedphysical phenomena.

1) Experiments of dehydration
    2) Precipitation tests
    3) Observation of crystalline Restructuring
    4) Atmospheric Analysis
    5) Analysis of flocculation Colloidal
    6) Magnetometer Study
    7) Biological Measurements
    8) 1st Experiment DETERMINING
    9) 2nd Experiment DETERMINING

Analysis of water treated with pyramid is a wonderful experience.

1) EXPERIMENTS TO PREVENT DEHYDRATION

There are placed two water sample of identical origin and equal in quantity in the pyramid and outside, away from your field but in the same conditions of temperature (in the same room). Looking for another result - difference of precipitation - in a recent experiment carried out with my friend and colleague Markus Jaime Salas, using sea water, in eighteen days the witness evaporated 50 cubic centimetres, while the pyramid evaporated 150 cc. . . . This phenomenon is due to the fact that the water molecules can be restructured in their proper form: pyramids of FIVE monomers of $H2O$, in such a way that all the monomers (proto-molecules of $H2O$) that are not balanced quantically complete to form molecules, are swept away by the magnetic field pyramid. The dehydration product of pyramid is that reorganization, by which the water that remains, is a little more difficult to evaporate at the end of experiment (about three weeks). Equal evaporates, but slower. This dehydration should not worry users of pyramids, already that we drink water and it simply is corrected, quickly closing the components that are not well structured.

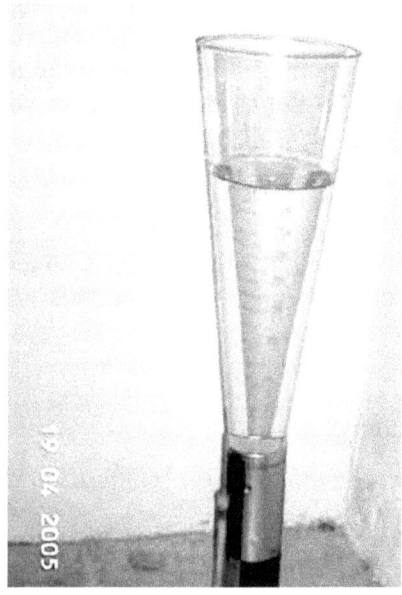

2) PRECIPITATION TESTS

The result in terms of precipitation (this not what we had done before), also gave their surprises. While the variation in the witness is a minimal increase, stable and distributed by salinity, proportional to the 50cc evaporated, in the pyramid we have a evaporation of 150 cc. (Three more times), with fluctuations of the surface salinity, centre and bottom of the beaker. This experiment can be done at home or in a laboratory elementary, relying only on pipette extraction and eyedroppers, distilled water and a hydrometer (optical, in our case). The results of the latest experiments have confirmed the theory of mechanics that makes the pyramid as an effective preventive consider antilithiasic.

3) CRYSTALLINE RESTRUCTURING

Similar to the previous high school, but it requires some more expensive items. Using a knife or razor blade USED, previously the microscope view (some 400 increases) and exposes it to the pyramid for a few days, after which you will be able to observe as the edge has been recovered, because the glass of the steel will have been reordered. This effect has two direct causes: Elimination of water proto-molecular in the interstices of the steel, of the microcrystal's tension activity steel. The latter can be seen with microscopes of maximum increase - some 2,000 - and subjecting the crystals to spectroscopic analysis (before and after treatment pyramidal). It clear that not only steel is used for these measurements probative.

Evidence of the formation of ice crystals in pyramids (using chemical freezing, better than electric refrigerators), is to hallucinate before the microscope...

If you have a scanning electron microscope, the number of tests possible increases to tens or hundreds of them, since all the macro impacts are caused by the quantum activity, atomic and molecular in the pyramid. Try to understand the effects of the pyramid without taking into account the properties and structure of authentic water [5 (H2O)] or equivalent H_{10} O_5 and the atomic constitution of the Universe and quantum, it is like trying to understand physically the wind without knowing the air.

4) ANALYSIS ASPIRATED

It is done with thermometers, barometers and hygrometers for accuracy. The evidence of possible control is several and only requires a stable environment and broad, with a pyramid of one to two meters of base oriented properly. it will be interesting differences, which left to find the experimenters (check at the same time below, outside, inside and above.

5) CONTROL OF COLLOIDAL FLOCULATION

You can do this any biochemist. As has been repeated countless times, for being one of the most important experiments from the point of view of the extension of cellular life, deserves an entire book, but very in synthesis: "delay" the beat SOL-GEL of the cytoplasm, lengthening the life of the cell, while decreasing the risk of alteration of the DNA. The gamma rays and Beta, main cause of the atrophy of the DNA, have their greatest moment of incidence during the process of change, from a solid to gelatinous (and vice versa) of the colloids of the cytoplasm and the nucleus. To this we must add the effect of elimination of free radicals, which we see below:

6) MAGNETOMETRY

Now we turn to something more complex and expensive, but many laboratories possess. With a set of magnetometers and ionic flow meters, you can check that:

a) The flow of negative ions is much faster in the pyramid that in any other body geometric compiled with the same materials and size (volume) approximate.

b) You can check that the amount of positive ions in the atmosphere is negligible and pyramid that are excluded from its field just enter in the. A percentage not determined cannot even enter the pyramid field, except that we drag when we walked in, which are then quickly removed.

7) BIOLOGICAL MEASUREMENTS:

You can make countless biological experiments, with seeds (which keep their germination almost indefinitely as long as they remain on the pyramid and retain this much then unpacking). There are also excellent averages of vitality, size and quality of plants whose seeds and/or seedlings are dealt with in pyramids. The variety of biological experiments possible is enormous, with crops of various bacteria, with houseflies, whose larvae will die of hunger even though they have sufficient samples of "lunch" (eases of any bug). The Roraima blue flies, bees, wasps, and not any insect larvae saprophytic draw a few plump, resistant and major.

8) 1st EXPERIMENT DETERMINANT for the purposes of the better utility

Convince yourself to a rheumatic very sick, as far as possible with osteoarthritis deforming (of which are harried medicines), to make them sleep a few months in a pyramid of aluminium from more than fifty Kg/m3, with sufficient angle wing to cover a 25% the total surface area (sides and base). Keep medicated as prescribed by a doctor, (only to relieve the pain, because there is no cure under allopathic medicine and will soon cease to take medication) and convince them that do not have to be afraid by the strange sensations that will have. After four or five months i account if the rheumatism cure or not.

9) 2nd EXPERIMENT DETERMINANT for the purposes of the better utility.
Get yourself some sick with intestinal infections several, full of bacteria by drinking infected water and that it has survived by the hair. Bake in a pyramid of sides closed, much more powerful than a normal Pyramid house; or at least with half of the surface cover, so that it stays there 18 hours a day. Range the only antibiotics attack and suspend gradually (with medical supervision). Two weeks before the analysis will be "rare" for biologists and will give you the definitive discharge before two months, without understanding the "rare residual effect of the

antibiotics". There is no residual effect. Simply the pyramid will be neutralizing the reproductive ability of the bacteria and the saprophytes interference until their extinction and without prejudice to the symbiotic bacteria of the intestinal flora. This happens because the pyramid energy prevents putrefaction but not digestion. These are two different things.

The best experiment is his own one, because the others are merely references, more or less valid. If you don't like to tell the movies and you prefer them, if you prefer listening to the music and not to comment, if you prefer a walk in the forest and the mountain more than seeing it on TV, if you prefer to make a trip in time to look at the photos of the travel of the neighbour ... What do you expect to experience what is theirs?

Chapter X
THERAPY WITH ANTI-PYRAMIDS MODE

You can use safely inside the pyramids of wood (except some very heavy and hard species), especially aluminum, glass and some plastics. You can stay all the time inside. But with the Anti-Pyramide, it is necessary to respect the times determined by doctors.

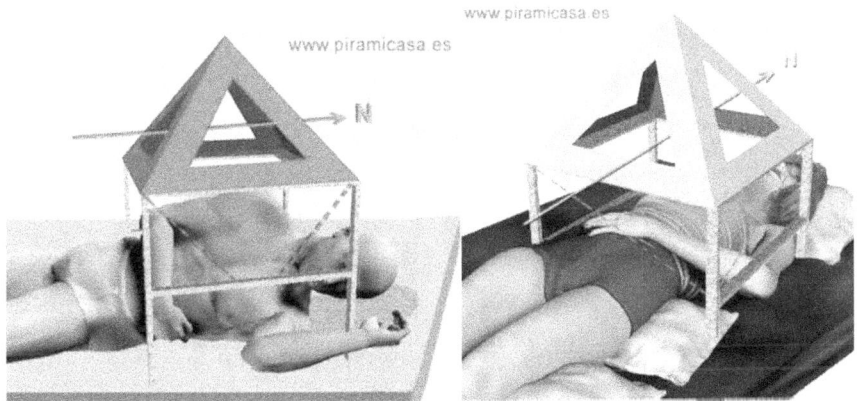

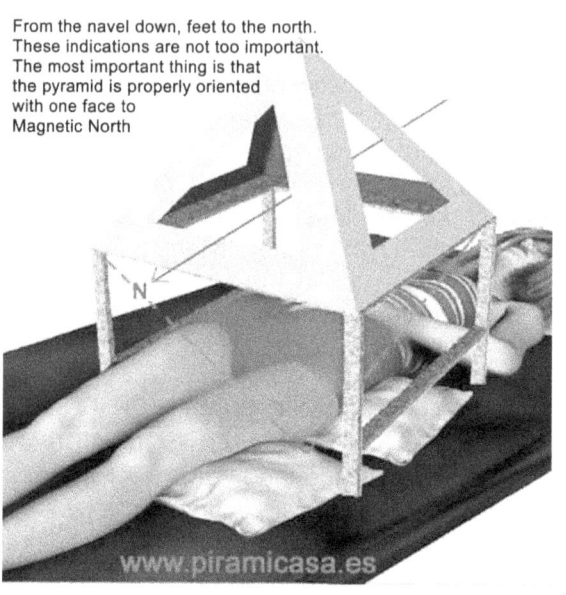

From the navel down, feet to the north. These indications are not too important. The most important thing is that the pyramid is properly oriented with one face to Magnetic North

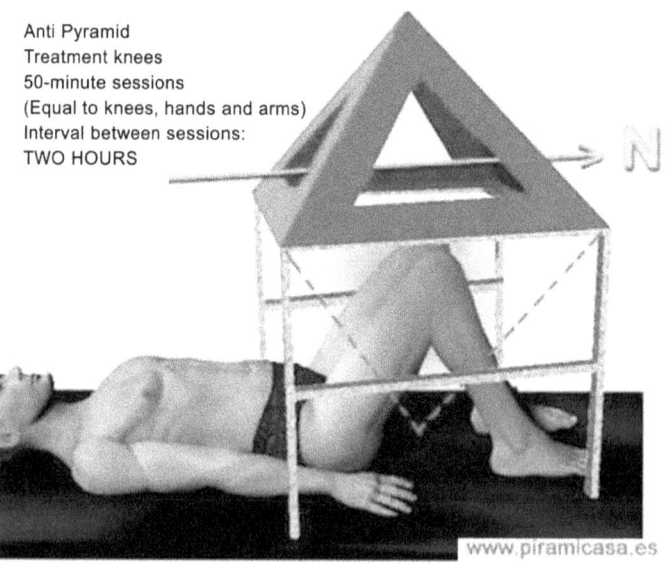

Anti Pyramid
Treatment knees
50-minute sessions
(Equal to knees, hands and arms)
Interval between sessions:
TWO HOURS

ANTI PYRAMID
Sinusitis, ear infections, dental infections,
eye and other organs of the head,
Hypo- and hyperthyroidism
Maximum Session: 10 minutes
Range: TWO HOURS

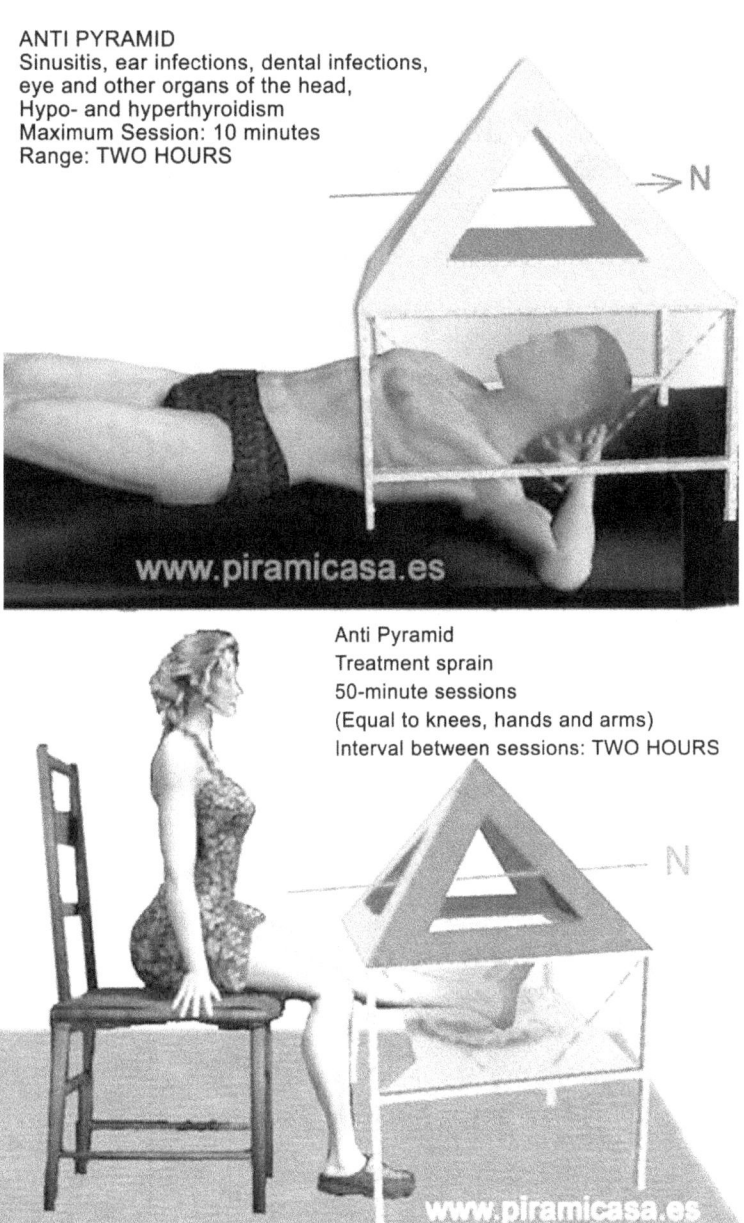

Anti Pyramid
Treatment sprain
50-minute sessions
(Equal to knees, hands and arms)
Interval between sessions: TWO HOURS

Anti Pyramid
Treatment hips
25 minute sessions
Interval between sessions:
  TWO HOURS

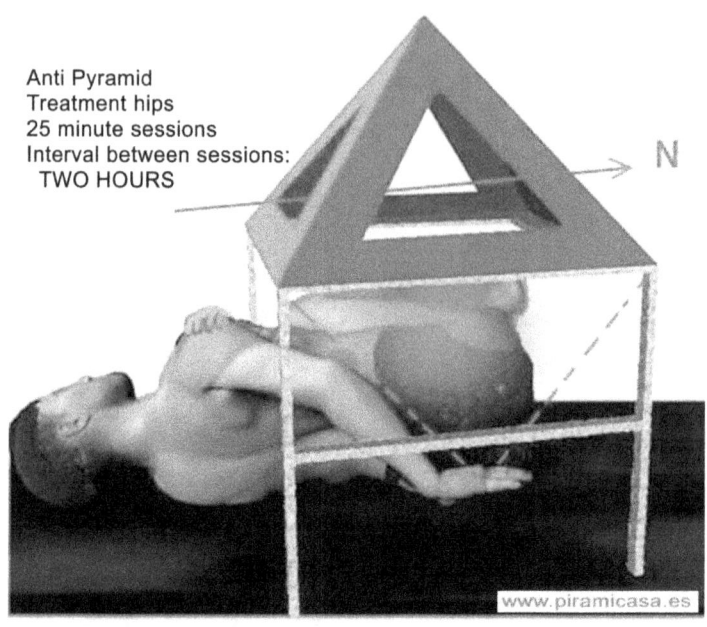

Anti Pyramid
Treatment arms
50-minute sessions
Interval between sessions:
  TWO HOURS

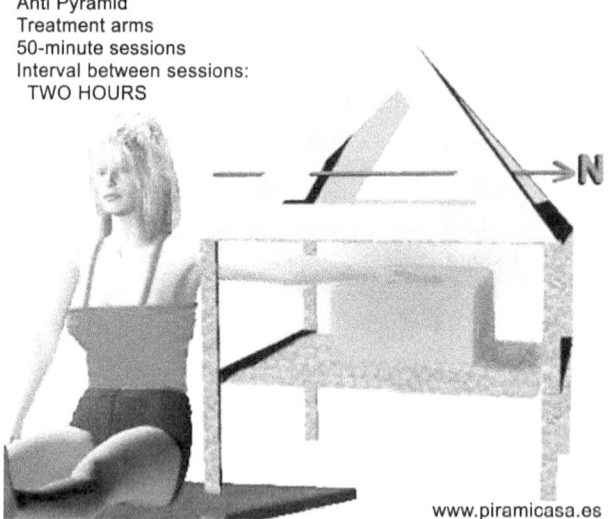

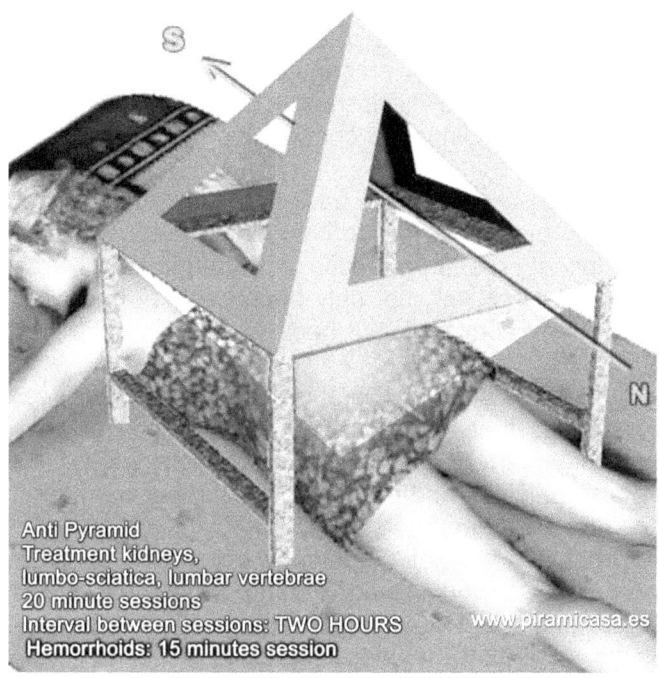

Anti Pyramid
Treatment kidneys, lumbo-sciatica, lumbar vertebrae
20 minute sessions
Interval between sessions: TWO HOURS
Hemorrhoids: 15 minutes session
www.piramicasa.es

Chapter XI
FRIENDS AND ENEMIES OF THE PYRAMIDOLOGY

The Pyramidology has friends and enemies, that for the experimenter an amateur as well as for the scientist it is important to know, in order to "separate the wheat from the tares" to the time to collate information and views:

ENEMIES OF PYRAMIDOLOGY:

The more powerful are the multinational economic interests of the pharmacopoeia, but at the same time the more worried about not to talk of them. The pyramids, like many other therapeutic elements, represent a very important alternative to hundreds of drugs, most of whom do not cure, can hardly alleviate some symptoms at the same time that produce plenty of collateral damage. But they represent billions of dollars a year in profits for a few companies.

On a second front (in importance but first to serve as cannon fodder for the economic interests), academic is the dictatorship that continues to sustain the absurd and unfounded idea of the pyramids in ancient tombs. These academic groups have their market in the sale of books for a "captive consumers" rep presented at students of archaeology, anthropology and related disciplines. Change things implies that would cease to assert their books and comfortable chairs and performing would be outdated.

They would be obliged to take back and apologize to others or to erase and re-write the history, as we do independent researchers. Fleeing from the interdisciplinary research, because any engineer or architect accepts his theories "constructive", not only because they have never consulted them in what they comment, but because none would ever work with a similar absurdities.

The petrologysts, chemical, and criminologists have ways to identify many more things that the age of a piece of stone or any material, but every time they participate, the archaeologists were fall the theories. There are also religious and political interests, but with less influence and serving the first mentioned.

A third front, by way of negation wing, which usually acts without conscience of those who served (though there are also very well-paid), are the false skeptics who only to denigrate the researchers, but that never would contribute nothing useful to anyone's life. To perform experiments to "show that it does not work" or that the pyramid does not "exist" the pyramid energy. Any person with its faculty's healthy brain knows perfectly well that prove that something does not work, it is the easiest thing in the world.

In the ala apparently contrary (but equal effects) are the pseudo-mystics, charlatans and mythomaniacs that distort reality, allocating to the pyramid incredible miracles, ability to transform a simple mortal in enlightened teacher of a day to another, drawn from the lottery by setting the number under the pyramid and absurd by the style.

It is true that the pyramids have relationship with the transcendence of humanity, that help or induce others to produce psychological catharsis, that prolong the life and perhaps we may find very pleasant surprises in the future, as we've had so far; but the pyramids that we now, according to the knowledge that we have only served to live better, to cure and prevent a broad spectrum of disease, apart from the accessories uses detailed above.

The fourth face of the enemy Pyramidology is the literature and cinema, which mixture tombs, terror and mummies with pyramids, creating a rejection in the minds of the public at large. Never found any mummy in any pyramid, but the people "reported" by the movies, rather than by the manuals or documentaries. Even most of the documentaries in this regard are riddled with contradictions and absurd theories exposed as proven truths, because archaeologists and Egyptologists, with such retain their chairs and the validity of their books, intended to make engineers and architects.

FRIENDS OF THE PYRAMIDOLOGY:

These are being added every day and I think they are already a force impossible to stop, because you would have to silence many thousands of people in the academic and practical millions of fans, experimenters and regular users of pyramids.

The first front is composed of medical doctors, physicists and experimenters in general.

There are doctors from around the world forming in pyramid therapy in Cuba, Nicaragua, Argentina, Uruguay and Mexico.

A pharmacist, biologists, microbiologists, physicians, patients and users in general, after their own experiences preserves and disseminates knowledge of this important civilizing tool. The effects are so easy to check, and with almost zero cost that it is impossible to "enemies" attack with basis against the primatologists.

Another front positive is composed by editors and journalists, along with young archaeologists of true vocation, whose intelligence does not allow them to accept the drivel that they are treated to teach. Also botanists, ecologists, farmers, ranchers and veterinarians are taking important benefit of the pyramids and when analyzing the impact that begin to take in our society and is inferred as increases, the view is very optimistic. A truly human society not only leverages the pyramids, but also that it deserves.

Currently (March of 2016) there are over 3112 people sleeping in beds pyramid. More than 3000 doctors use pyramids in Cuba, about five hundred medics in Spain and more than a thousand elsewhere. The "Pyramid Revolution" is unstoppable, but not as a fad, but for extraordinary anti-aging effects, anti bacteria, the effect of molecular restructuring, antirheumatic effects, as medicine anti-sclerotic non invasive... And much more...

SYNTHETICALLY: The general effects of a Perfect Pyramid are products of two fisic phenomenons: The magnetic field in the making, of high frecuency no electric, and the acumulation of neutrines, who have benefic propertys for all superior organisms. This effects are produced for the called "pyramid energy".

THE BEST HEALTH INSURANCE

The best Healt Insurandce is not sick, to have the body "armored" against degenerative, infectious, rheumatic diseases, and sclerotic.

Anti-aging Effect anti-oxidative without chemical products.

Anti-rheumatism Re-estructuration molecular of the water and all liquids components of sinovial capsules,

Bacteriostatic Prevents any process of putrefaction.

Anti-litiasic Maintains suspended colloids, preventing the formation of crystals and stones in the organs.

Relaxative Produces magnetic balance. (biomagnetic pair) naturally. It tones the nervous system.

Anti-sclerotic, correction of homeostasis and much more.

The pyramid es an instrument of big curative power for many diseases, but the oficial medicine have not a serious researchs, except in Cuba. But many Doctor of Medicine, way personal and professional secret mode, reach excelent result in diverse pyramidal terapys,

particularly with the rheumatics diseases (all aetiologys), digestives and sleep disturbances, infections, etc., but they don't like make a comment on their results for fear to the derision or to make a fool before their colleagues. The Pyramid cause dehidratation, but force to body, for quicken the natural level of thirst. This effect is extraordinarily benefic for the organism, because whom live at the Pyramidal house need drink more water.

Besides, the spontaneous evaporation for magnetic effect, produde the elimination of the organic excessive or not interactive liquids, as which they cause the drop, excesses of gastric liquids, etc.. Several hundreds of persons in the whole world, know already the pyramidal effects. Also it has reported of several cases of slimming without any side effect, so that it is effective against the obesity. The water that it remains in the experimental pyramids for fifteen days, he acquires corrective properties of diverse intestinal and gastric problems, as well as it turns out to be a stimulant of the processes of treatment of wounds and sores, so much like internal day pupils. The pyramidal houses possess a special water tank for this purpose. His more notable effects in the psychological thing, are the antidepressants and relaxing. The dream is a deeper and better opportunist. In general, there can be deduced the great quantity of benefits that it bears to remain in an environment without radical free, (Which delays the process of aging, since he avoids to a great extent the cellular oxidation). Saturated of neutrinos, (that revitalizes the organic active matter) and where the magnetic current reorientates the molecular structures, correcting affections derived from the distortions of the above mentioned structures.

All of these benefits are also applicable in veterinary.

People, dogs, cats, horses, cows, pigs, chickens, ducks, birds in general... And all vegetables can live or cure in the perfect pyramid. Only the saprophytes can not live in the pyramidal atmosphere. The general effects of a Perfect Pyramid are the products of two physical phenomenons: The magnetic field production of non-electric high frequency and the accumulation of neutrons that have beneficial properties for all superior organisms. These effects are produced for the so-called "pyramid energy."

This dog was poisoned with strychnine. The vet said it was not possible recovery. You may not always be effective. But many ailments treated animals. The animals receive the same benefits. The chemical effects of the pyramid are derived from the strictly physical effects. And it is a restorative effect on the quantum, atomic and molecular level, therefore no chemical conflict. Its utility has the same parameters for plants, animals or people.

These puppies were born in 2008 and living in pyramid. In the picture below they are in your current age (March 2016). They were never vaccinated. They have never had nor have any disease.

Working with bee. Cuba was eradicated in the ascosferosis and varroa, using large and small pyramids.

HEALTH BENEFITS AND USE (UTILITIES) OF PYRAMID POWER

Antiaging - An anti-oxidative effect without chemical products

Anti-rheumatism - molecular restructuring of the water and all liquid components of the synovial capsules

Bacteriostatic - prevents any process of putrefaction

Anti-litiasic - maintains suspended colloids, preventing the formation of crystals and stones in the organs

Relaxative - Produces magnetic balance (bio-magnetic pair) naturally. It tones the nervous system.

Besides the spontaneous evaporation for the magnetic effect, it produces the elimination of organic excess or non-interactive liquids which cause the drop or excesses of gastric liquids, etc. Several hundreds of people worldwide already recognize the pyramidal effects. There have also been several cases of weight loss without any side effect, so it is effective against obesity. The water that remains after experimenting the pyramid for fifteen days acquires corrective properties for diverse intestinal and gastric problems and it turns out to be a stimulant in the processes of treating wounds and sores as well. The pyramidal houses possess a special water tank for this purpose. The more notable effects are psychological, antidepressant and relaxing. The dream is a deeper and better opportunist. In general, a great amount of benefits can result which remain in an environment that is radical free; which, in turn, delays the process of aging since it avoids to a great extent the cellular oxidation. It is saturated with neutrines which revitalize active organic matter where the magnetic current redirects the molecular structures, correcting affections derived from distortions to the above-mentioned structures.

Hercules pyramd, very powerful for treated multiple sclerosis, rheumatism and other diseases.

NOURISHMENT:

In food the same qualities acquiring revitalization from water and remain extraordinary. The lacteal ones present variations in their condition (state), especially in yogurts where they remain so a longer time. Meats do not rot and, the first days without any refrigeration normally a thin cap forming "penicillin" which later remains inert (passive) while the interior is mummified. When it returns to the normal environment, it usually rehydrates (not always) in a few minutes spontaneously with the environmental dampness.

Cut vegetables are kept fresh equally to their time in the normal environment. Some of them, specifically those of leaf) are dehydrated quicker, but instead of losing flavor, they remain tastier to the palate. If you leave them in water, they keep longer and do not rot. Honey becomes thicker, even crystalizes, but will usually spontaneously return to its gelatinous condition in changeable and alternative circumstances. The taste is intensified. Wines, tobacco, liquors also improve their qualities of aroma and flavor.

AGRICULTURE:

One of the most important practical applications of the pyramids is in agriculture. It is the quality of preserving the normal thing longer; the power of the seeds to germinate. Besides this power being more outstanding it increases between 20 and 30 percent according to the variety. In pyramidal seedling, much insures a healthier and more fruitful production. There are also improved effects of certain chemical products, but our recommendation is to not use poisons, but to do mixed cultivation; planting fragrant varieties (oregano, bay, lemon grass, etc.) in between other produce. The water treated in pyramidal warehouses (deposits) can give results that will amaze whoever experiments with it.

It must be taken into account that the pyramid is not useful to attack insects, since they are top organisms their larva is usually larger and healthier. On the other hand, insects (especially mosquitos) can be kept away from the pyramids by painting the frames of its doors a strong color or a violet lilac color. Although this may look like nonsense, it is not and the reader can apply it to his house and prove it.

A good idea of experimentation would be to use a pyramidal beehive because the bees would be very benefitted and the homey would turn out to be extraordinary. Until now we have not been informed about experiments in this field.

Seeds treated in a pyramid last permanently and their viability is retained indefinitely.

The benefits and usefulness of the pyramids are innumerable and extraordinary since it will be understood that they are not simply a fine line, but incredibly cover the systems of interests in which we live, supporting the ignorance of those things than provoke a revolution in humanity's way of life. As well as preventing the development of ways to extract ecological energy, many advancements hide themselves because they turn out to be disadvantages to marketing interests.

Characteristics:

Pyramid closed for seed conservation (Very poweful)

This model has from 80 cm to any size. Structure of pure aluminum with cover of aluminium too, but have a small door. You don't need move the pyramid for work inside. The base is aluminum or wood depending on use, location and type of seed to be treated.

Still, complete research could not have been done on the extraction of electrical power by application of pyramidal ingenuities, but the principle has been demonstrated and is possible. We will continue adding material as it becomes available.

The most important use today, is to sleep in a pyramid. With the same effects that sleep in the King's Chamber of the Great Pyramid of Giza.

For agricultural uses in large plantations, you need to plan the strategy. Plants should be treated before planting, but it is also possible to treat trees have already been planted.

One of the most common uses is to treat irrigation water. In that case special pyramids are used, but with different materials. No need to use extreme purity aluminum, but we need to pyramids of great power. However, these pyramids should be used with care, because they are not suitable for people or animals are within them.

We also use smaller pyramids, but powerful to treat pests of trees. In some cases you have to try the roots, in other cases be treated foliage.

For treatment of liquids such as oil, wine, water, fuel, liquor, chemicals in general, we can use ferromagnetic pyramids. But our recommendation is not to use diamagnetic pyramids in any case. Copper pyramids are especially dangerous.

Some specialized in the production of flowers, pyramids farmers use since 1960. Also the pyramids have been important in agricultural development in Cuba. In the Netherlands, Finland, Romania and Russia, some projects have been funded by States. But economic interests have always sabotaged such projects. However, many private companies use the pyramids, hiding the results. For them it is an important secret, because they want to maintain a monopoly on knowledge. That's one of the reasons that the Pyramidology no more known to the public.

PYRAMIDS AND TOURISM

The future hotels, like houses, hospitals, and most of the buildings in the world, will pyramidal shape.

The Luxor Hotel of Las Vegas, is already demonstrating the wonderful benefits of pyramidal shape. Not talking about a simple aesthetic issue. It is organic effects, "feel good" and better as more time we spend inside the pyramid.

The investors who have good vision, build the first Pyramidal Resorts, as "Villa Piramidal", in México. The "health tourism" is a modality that is growing every day, and no construction has many benefits as the pyramids. Apart from the holistic, alternative and innovative medical treatment it is healing just because sleeping inside a pyramid.

CHAPTER XI – THE MAXIMUM SYMBOL OF HEALTH

Anyone, anywhere in the world knows that the "Cross of Pharmacy" is the global symbol of health. But few people know that this symbol is a pyramid. Yes, a pyramid inscribed in a plane. The Cross of Order of Malta has broken sides inward. This is a symbolic representation of 27 minutes of angle in the laterals of the Great Pyramid o Gizhé.

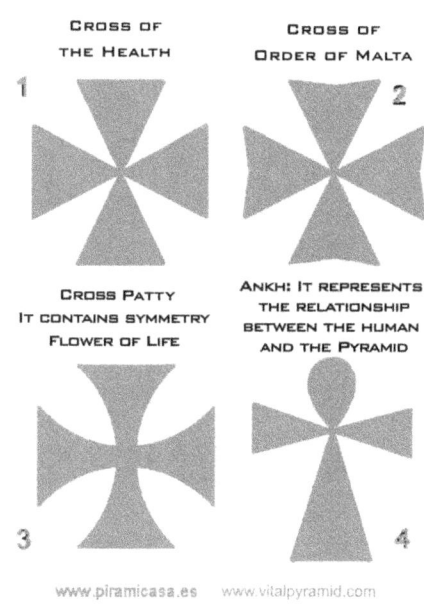

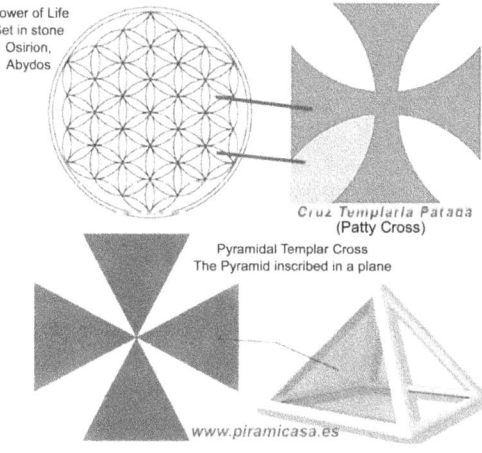

The symbols have always been at the sight of everyone. But almost everyone is amazed to understand its meaning. The union of the Pyramid and the Flower of Life, constitute the esoteric symbol of the most famous Templar Order. But even most of the Templars (of the Order of Malta and almost all) are surprised, because they do not know that the pyramid is much more than just a symbol. The next image is a doctor's office in Nicaragua: www.auroracentro.com

CHAPTER XII – THE GREAT PYRAMIDS IN THE WORLD

Regarding the ancient pyramids, we are rewriting history. It is not easy work, but we're doing.

Here is a revelation of our findings: It is a lie that the hieroglyphs have been translated. Our research team has translated only 28 pure hieroglyphic and 30 indicative. They have nothing to do with the demotic language, which is a little more and some reliable translations.

This is the authentic name of Egypt: *Ankh em-Ptah*. Ankh is the Key of the Life. Ptah is "*Divine Essence*" depicted as a bearded baby, because it is always born. He has a cap swimmer, because it is in "the Primordial Waters." And he has in his hands the Uas (kind of powerful weapon), Heka (magic), Nejej (sort of divining rod). And the Ankh attached to Djed (a type of capacitor).

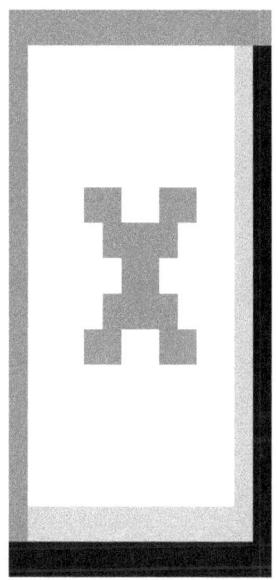

When I was not yet six years, I realized that the Great Pyramid was not a tomb. That great work could not be a tomb. My father was an engineer of great works. He was surprised when I told him that the pyramid could only be a device, something that worked somehow ... There was no reason to assume such a complicated work, only serve to bury someone.

Since my thirteen year old I started practical experience, I built my first pyramid (such as the template). And after fourteen trips to Egypt, I am convinced that everything, absolutely everything there is in ancient Egypt, is product of more advanced than our technology.

Next image: "Ank em-Ptah"

But the hieroglyphic most repeated all over Egypt, in the oldest temples and the most modern, is this: *Ankh em-Pyr*. "The Key of the Life is in the Pyramid".

No need to be a rocket scientist to realize this. It's just a matter of common sense reasoning. But for the official archeology, it is more important to preserve their chairs, their economic interests, take care of their "captive customers" (students of archeology) destined to repeat their mistakes and deceptions. Therefore it is necessary to rewrite the story.

Pyramid of Olmedilla, Spain
Found that it is artificial
300 meters of diameter

The ancient builders, who gave us great pyramids worldwide, knew the value of Pi, Fi, the structure of the water molecule (a pyramid consisting of five monomers H_2O). They were not ignorant pharaohs bow and arrow. These Great Builders (much more older than the imitators pharaohs of recent dynasties) knew quantum physics, higher mathematics, geometry, astronomy, geodesy, geology, chemistry, superior engineering... And all scientific disciplines, with more depth than today's scientists.

Pyramid of Bonete, Spain

Pyramid of Cañete (Cuenca, Spain)

Is so long and wide the theme of the ancient pyramids, I prefer to let readers find more information in videos and history books (not the official books). These small pieces of information are key to that search.

You can start up a stage of change of criterion about the Story, but searching a True Story... Egypt is the largest center of that ancient civilization, but it is not the only place in the world where there are great pyramids. This webs era a "Big Door" for start your personal search.

www.vitalpyramid.com   www.piramicama.com   www.piramicasa.es
www.piramicasa.com   www.piramidedecuenca.com   (pyramids in Spain)
www.bosnianpyramids.org

A marvelous documental about the ancient pyramidal civilization:
https://youtu.be/cROmKEXESnM

This is your new bed...? Only sleeping or living in a well constructed and oriented paramagnetic pyramid, you will find the best and deepest secrets of the Pyramids.

This may be your new house. Nothing rots in the interior…
We also use highly advanced energy technology, exclusive to users of Pyramid-houses

Write to Gabriel Silva – Piramicasa - piramicasa@gmail.com

www.piramicasa.es - www.vitalpyramid.com

www.ingramcontent.com/pod-product-compliance
Lightning Source LLC
Chambersburg PA
CBHW070429180526
45158CB00017B/944